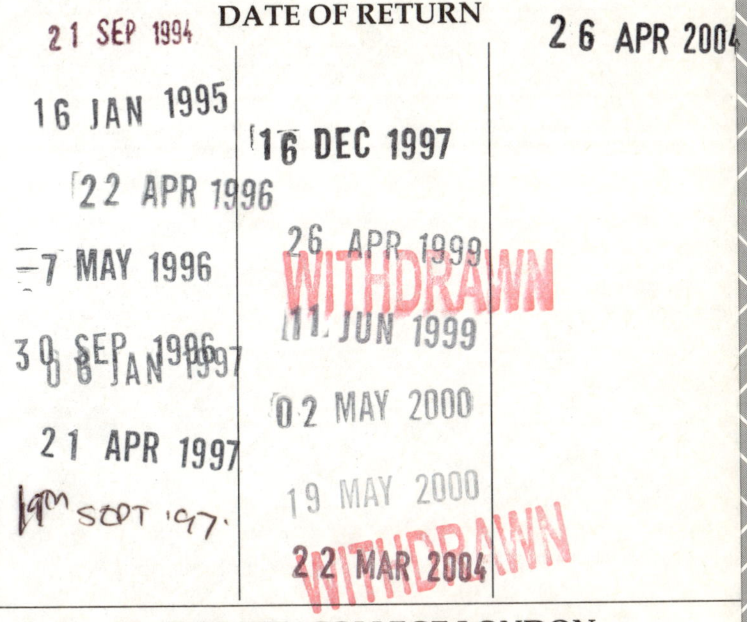

This item must be returned or renewed by the last date shown below. The loan period may be shortened if it is reserved by another reader. A fine will be due if it is not returned on time.

DATE OF RETURN

21 SEP 1994

16 JAN 1995

22 APR 1996

7 MAY 1996

30 SEP 1996

08 JAN 1997

21 APR 1997

19th SEPT '97.

16 DEC 1997

26 APR 1999

WITHDRAWN

11 JUN 1999

02 MAY 2000

19 MAY 2000

22 MAR 2004

WITHDRAWN

26 APR 2004

UNIVERSITY COLLEGE LONDON

Gower Street London WC1E 6BT

LF4D

Managers and Innovation

In the management of innovation the key resource is the firm's knowledge base. Managers' understanding and interpretation of this base constrains how they select and approach innovatory projects, but managers' experience of successive innovatory projects alters the knowledge base of the firm. The vital decisions concern which new technologies to develop, when to develop them and why they make sense in terms of the existing knowledge base and the managers' understanding of potential markets.

Using a detailed case study, the author gives valuable information about how managers have interpreted market, technical, competitive and other influences on their projects. The problems which arise from making technical innovation a viable business proposition are then able to be addressed. It is this process of innovation management on which the book centres. By way of analysis of the process of technical innovation in early biotechnology, the interplay of cognitive and social processes in the construction of new technology is shown.

This book is a source of previously unavailable data and should be read by students of technology management, as well as managers with an interest in biotech policy and industry.

John Howells is a lecturer on the management of technology and innovation at Brunel University, west London.

Managers and Innovation

Strategies for a biotechnology

John Howells

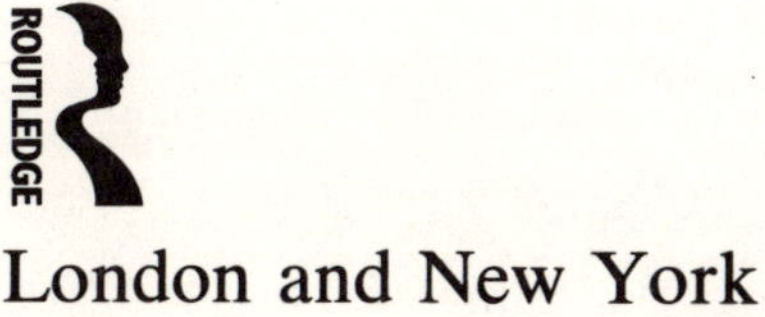

London and New York

First published 1994
by Routledge
11 New Fetter Lane, London EC4P 4EE

Simultaneously published in the USA and Canada
by Routledge
29 West 35th Street, New York, NY 10001

© 1994 John Howells

Typeset in Times by J&L Composition Ltd, Filey, North Yorkshire
Printed and bound in Great Britain by
TJ Press (Padstow) Ltd, Padstow, Cornwall

British Library Cataloguing in Publication Data
A catalogue record for this book is available from the British Library

Library of Congress Cataloging in Publication Data
has been applied for

ISBN 0–415–08590–X

To my mother and father

Contents

List of figures   viii
List of tables   ix
List of boxes   x
Preface   xi
Acknowledgements   xii
List of abbreviations   xiii

1 **Technical innovation and the economy**   1

2 **Introduction to the technology**   17

3 **The markets and the environment during SCP development**   40

4 **Cooperation and competition between companies**   71

5 **Firm and industry culture and the role of senior management**   82

6 **Technical choices in the development of the novel fermentation projects**   129

7 **Project champions and project patrons**   166

8 **A socio-cognitive approach to innovation**   180

Appendix   186
Glossary   211
Notes   213
Bibliography   216
Index   225

Figures

2.1 Flowsheet of Pruteen process 30
2.2 Summary chart of test procedures 32
2.3 Prices of methanol and crude oil 35
2.4 Prices of maize and crude oil 36
6.1 The relations between product and market, production technology and firm expressed within the key social division between the firm and the market. Innovation involves selective broaching of the firm–market division of knowledge 164
A.1 Flowsheet of the Cellulose Attisholz pulp-processing plant 201

Tables

1.1 Kondratiev waves and associated technical innovation 3
2.1 Company, product and approximate size of plant built 22
2.2 Substrate, organism and company 25
2.3 Percentage of protein in bacteria, yeast and fungi 28
2.4 Amino acid profile of common animal feed protein sources 31
2.5 Operating costs of 100,000 t/a SCP processes based on methanol 34
2.6 Capital costs of the Pruteen process 35
A.1 Cost of Bioprotein compared to cost of dried milk powder 186
A.2 Economics of Pekilo protein production 198

Boxes

1.1 A taxonomy of innovation   5
2.1 The principal substrates   26
2.2 Main events in the BP story   37
2.3 Main events in the ICI story   37
2.4 Main events in the RHM story   38
2.5 Main events in the Liquichimica story   39
5.1 Value of R & D in different companies   127
5.2 Culture or style linked to organisational sub-groups   128
7.1 Summary points on project patrons and champions   178
A.1 Summary of SCP development events in Japan   196
A.2 Principal stages of the Cellulose Attisholz yeast
     production process   203
A.3 Development of Bel Industrie's food yeast production
     process   205

Preface

This book is about the management of technical innovation. It uses a case study of radical innovation to gain insight into how managers interpreted market, technical, competitive and other influences on their projects, and analysis of this process of innovation is the focus of the book.

The case study consists of the attempts to make new sources of food available for animal and human consumption through bulk fermentation of hydrocarbons and other substances, the products of which are often called single-celled proteins (SCP). Almost every large chemical and food company was involved in this technology to some degree and at some stage during the 1960s and 1970s.

The internal rationales for entering this field were varied and led to surprising variety in the construction of the fermentation technology. This variety ensured that when economic shocks drove most projects to 'extinction' (a fascinating process in itself) a few survived. This book is, then, a case study in innovation failure.

Acknowledgements

I would like to thank all the interviewees who made this work possible. Without their time, courtesy and cooperation there would have been no book. Thanks also go to the many friends and colleagues who have given time to discussing this work.

Abbreviations

BOD	Biochemical oxygen demand
BP	British Petroleum plc
BMFT	West German Technology Ministry
DNA	Deoxyribonucleic acid
ICI	Imperial Chemical Industries plc
IFP	Institut Français de Pétrole – French Petroleum Institute
LQ	Liquichimica
MAFF	Ministry for Agriculture, Fisheries and Food
RHM	Rank Hovis McDougall plc
RNA	Ribonucleic acid
SCP	Single-cell protein
t/a	Tonnes per annum

1 Technical innovation and the economy

This chapter explains why technical innovation is important as a source of change and development in the economic long term. It argues that case studies of technical innovation within the firm can contribute to an improved understanding of how technical change occurs in the economy. There is then a selective review of innovation case studies, and some conclusions are drawn about how these might be improved upon. The chapter ends by outlining the 'conceptual framework', a collection of possible innovation influences that served to guide the initial interviews.

THE EXISTENCE OF LONG WAVES IN THE WORLD ECONOMY

World economic development among the industrialised countries has not been a smooth and continuous process, but rather periods of economic growth have alternated with periods of slower growth. Prices and employment indices have also shown long-term variations, with an apparent cycle of approximately 50 years between peaks in the indices (van Duijn 1983). These variations are often referred to as 'Kondratiev' waves, after the Russian economist of the same name, who first drew attention to them. There have been a number of attempts to explain their existence, the explanations having in common that the 'cause' of the long waves must vary on a time scale of approximately 50 years. It is possible to favour as candidates the changing structure of capital investment and long-term shifts in demand (Delbeke 1984), but those that are of the greatest interest here emphasise technical change as the source of long-wave variations and can be termed 'structuralist' explanations or models. In a sense the effort of choosing among such competing explanations is a false one because, as should become clear by the end of the book, when analysed at the micro level, 'technical' change is difficult to separate from shifts in the associated idea of a market for the innovation.

Almost all the structuralist models refer to Schumpeter (1942) as the first writer to link such waves to technical and structural change. One of Schumpeter's most frequently quoted passages suggests that he made this link because of his observation of the dynamics of firm competition in his contemporary economy. According to Schumpeter the most powerful type of firm competition was

> the kind of competition from the new commodity, the new technology, the new source of supply, the new type of organisation . . . competition which commands a decisive cost or quality advantage and which strikes not at the margins of the profits and the outputs of existing firms but at their foundations and very lives.
>
> (Schumpeter 1942: 84)

Schumpeter contrasted this powerful form of technical and organisational competition with the interests of mainstream economists, which were essentially interests in static forms of competition, 'competition within a rigid pattern of invariant conditions, methods of production and forms of industrial organisation . . . that practically monopolises [economists'] attention' (Schumpeter 1942: 84].

Long waves are therefore associated with periods where more vigorous exploitation of technical and structural change is taking place in the economy. A study of the variations in indices for economic growth, employment and investment since the Industrial Revolution prompts the identification of four distinct waves corresponding to periods of development. Freeman and Perez have associated each of these 'Kondratievs' with different clusters of technical and organisational innovations; a simplified list is shown in Table 1.1.

The waves occur as the aggregate result of many individual firm decisions to exploit the particular cluster of innovations that come to characterise a wave. In Schumpeter's model, entrepreneurs would reap temporary monopoly profits from successful innovation, and these high profits would attract new entrant firms and entrepreneurs to the same line of business. The increased competition from the new entrants would gradually reduce the high profits of early entrants and ensure the wider distribution of the financial benefits of the innovation. This pattern of increasing exploitation and eventual exhaustion of the commercial opportunities linked to a wave's cluster of innovations results in the early and late phases of each wave being

Table 1.1 Kondratiev waves and associated technical innovation

Kondratiev	*Typical innovations*	*Leading country*
Early mechanisation Kondratiev 1770s to 1830s	Textiles Textile chemicals Textile machinery Water power Potteries	UK France Belgium
Steam power and railways Kondratiev 1830s to 1880s	Steam engines Steamships Machine tools Iron Railway equipment Railways	UK France Belgium Germany USA
Electrical and heavy engineering Kondratiev 1880s to 1930s	Electrical machinery Heavy engineering Steel ships Heavy chemicals Synthetic dyestuffs Electricity supply and distribution	Germany USA UK France Belgium
Fordist mass production Kondratiev 1930s to 1980s	Automobiles Trucks Armaments Aircraft Consumer durables Process plant Synthetic materials Highways	USA Germany Other EC Japan
Information and communication Kondratiev 1980s to ?	Computers Electronic capital goods Software Telecommunications Optical fibres Robotics FMS Ceramics Satellites	Japan Germany Sweden Other EC

Source: Freeman and Perez 1988

associated with different employment and investment situations.[1] In the early phase investment is directed towards expanding capacity, so that greater profits can be taken from greater output, while in the late phase investment is directed towards cutting costs and shedding labour as part of an effort to compete on price.

An example of a 'cluster' of innovations which Freeman associates with the fourth Kondratiev are those involved in the production of synthetic oil-based materials (Freeman 1982). The plastics and synthetic fibres innovations were aided by the establishment of basic polymer chemistry which occurred in the 1920s,[2] but the industry in its modern form was only possible once certain high-pressure chemical engineering innovations (materials and techniques) were developed. These were essential to enable the construction of experimental and later full-scale synthetics materials plant, which, with their economies of scale, brought the unit price per tonne of synthetics so low that they were able to compete in certain markets with natural materials. In short, commercial synthetics were only possible once a range of technical and scientific innovations were established and brought together by innovating firms.

In his later work Schumpeter recognised that an increasing proportion of innovations were conceived and developed in the corporate R & D department.[3] The R & D department is one of the major industrial organisation innovations of the third Kondratiev, and it has continued to diffuse through certain industrial sectors during the fourth Kondratiev (Freeman 1982). The growth of the industrial R & D laboratory has been accompanied by the increased importance of some types of science as an input into innovation.

Of course, not all technical innovation is equivalent in the scale of investment it requires or in its market impact, and it is possible to develop a variety of qualitative innovation 'taxonomies' to represent this idea. In their innovation taxonomy, Freeman and Perez stress the interaction of technical and non-technical innovations as the base of each period of economic restructuring by labelling this type of innovation a 'techno-economic paradigm' (see Box 1.1). A techno-economic paradigm is the set of technical and organisational innovations which underlie each Kondratiev wave; the idea of a paradigm appears to be useful in emphasising the looseness of the assembly of innovations and helping us to resist the idea that any particular wave is necessarily and always related to the same cluster of innovations.

We do not have to take too seriously the idea that Kondratiev waves occur in cycles with fixed periodicity – it is difficult to suggest a mechanism whereby the downswing of one wave necessarily leads to the upswing of the next at a precise point in time. But the core idea that there are periods when the economy undergoes relatively rapid restructuring under the impact of the development and diffusion of a large number of innovations focuses our attention on the role of the firm and, within the firm, on individual managers, since

Box 1.1 A taxonomy of innovation

Incremental These are innovations which modify a part of a production process. They are common in all branches of mechanical engineering and often result from individuals working on the production process. They are limited in their economic impact, although their cumulative impact may be large in some industries, such as those based on chemical and refining plant.

Radical These innovations may be new products or production processes, examples being the invention of individual synthetic fibres. They require substantial investment to bring the innovation to market and their economic impact may be considerable but is localised. An example would be nylon, which radically altered the market for natural fibres for selected products, but did not enter into a substantial proportion of the total of manufactured goods.

Change in technical system These innovations can include organisational or management innovation and change the technical basis for production in an industry or for a significant range of products, but fall short of changing the entire economic structure. An example would be the collection of synthetic fibre innovations and its impact on the textile production and manufacturing industry.

Change in techno-economic paradigm These innovations have the potential to change the basis for production throughout the economy. They may require vast investment in a new infrastructure to support the new economic base. Electricity is one example.

Source: (Freeman and Perez 1988)

ultimately innovations are developed and diffused by the efforts of people working within the firm.

THE INNOVATING FIRM

If we continue to focus on the role of technical innovation, then a promising attempt to integrate technology and technical innovation into a model of firm behaviour uses the properties of technical knowledge to identify the firm itself (Teece 1988). According to Teece, 'A firm's core business . . . stems from the underlying natural trajectory embedded in the firm's knowledge base' (Teece 1988: 264).

In this model, firms specialise in certain products because they possess the requisite technical knowledge base. This knowledge base constrains the range of possible innovations, but succeeding innovations shift the knowledge of the firm so that it can be thought of as moving along a trajectory in time, the trajectory being defined by succeeding generations of production technology and products.

One of the values of depicting technology as a form of knowledge is in explaining aspects of firm behaviour. Technical knowledge is characterised by its tacit nature;[4] it exists only partly as formal, written work, but substantially as 'practical' knowledge possessed by technicians, engineers and scientists. The tacit nature of technical knowledge results in the difficulty of transfer of skills and knowledge between firms without a transfer of personnel. Firms will tend to want to stay within their core business based on proprietary technical knowledge, because of the costs and difficulties of changing their technical knowledge base.

Technical innovation is deliberate in nature and the technical base of the firm does change through time as technical innovation is deliberately introduced. The idea that the firm's competitive decisions are responsible for its introduction of technical innovations and that it is these innovations that mould the development of their technical base, is emphasised by Metcalfe (1988).

> Technologies do not compare in the literal sense. Only firms compete, and they do so as decision-making organisations articulating a technology to achieve specific objectives within a specific environment. The outcome of their decisions is what determines the economic significance of rival technologies and how this changes over time.
>
> (Metcalfe 1988: 568)

This gives the sense that technologies come alive in the minds of the managers of the firm, and that by their existence they express past competitive intentions and guide future competitive decisions.

THE BEHAVIOUR OF MANAGERS

As far as there are similarities in the trajectories charted by individual firms then a pattern at the macro level arises in the kind of technological development choices which justifies the concept of 'technical system' or 'techno-economic paradigm'. That such patterns exist on the macro level from a retrospective view should not blind us to the uncertainty associated with decision-making within the firm.

One of the features of decision-making that is frequently noted by authors looking at how and why decisions are made, is the rule-of-thumb.[5] For example, Kay infers from differences in R & D spending between firms that there are 'meta-rules' suitable for each corporation's circumstances that managers use to set levels of spending on innovation (Kay 1979a). These rules are based on experience and are disembodied from complex, 'rational' calculations. Their advantage is that, based on past experience, they do not require time and effort to generate each time individuals are called upon to act. Elsewhere, in a review of the budgeting and project selection literature, Winkofsky and Mason come to similar conclusions about the value of rules of thumb, or heuristics. In their view, project selection is not a 'constrained optimisation problem', as is commonly thought, but rather, it is a 'highly diffuse and heuristic process carried out by many individuals and groups within a firm' (Winkofsky and Mason 1980: 12).

This contrasts with what might be termed a simple 'rational' view of decision-making where decisions occur at well-defined points in time and represent optimal choices based on full access to information.[6] The problem that decisions often do not appear 'rational' in practice is highlighted by the following quotes from a PhD thesis, where the author used participant observation to compare in detail some of the characteristics of two electronics companies' R & D departments. He aimed to, 'assess [the] information and economic nccds of dccision makcrs so that thc dccision-making function could be reduced to a logical evaluation of alternative outcomes' (Thomas 1970: iii). However, after more than two years' work Thomas concludes that 'The rational planner decision-maker models are impracticable' (Thomas 1970: 598).

It is significant that empirical studies of manager behaviour, such as the above, frequently question the value of the rational planner model, and rational planning techniques; this suggests that these techniques and the model itself are of far less value than is commonly assumed. If we ask what we should use in their place, a guide is Gold's characterisation of management decisions as 'elements in a stream of temporary successive commitments' (Gold 1971: 22). Rather than lose the idea of rational behaviour completely, we can think of managers having a 'bounded rationality' (Cyert and March 1963) where, within their assumptions and beliefs and with limited access to information, management decisions do make sense and are 'rational'. This still leaves the problem of how to conceive of an organisation if the individuals within it have bounded rationality, incomplete access to information and use heuristics to make decisions.

Change in the organisational environment within which managers operate is a regular occurrence, and individuals in industrial organisations have constantly fluxing attitudes in response to change. When faced with major organisational crises there is a reorientation of their attitudes and beliefs, with the reorientation being conditioned by prior experience – this has been called a 'processual perspective', because there is a process of negotiation of organisational reality which continues between all levels and individuals in an organisation (Elger 1975). This perspective emphasises the negotiable status of the meaning of changes in the firm environment and external influences such as new technology. Hence, despite such apparent external constraints, there is a degree of choice exercised by managers through their interpretation of the meaning of such change.

Within this perspective managers should be treated as 'theorists' in their own right because they strive to make sense of their organisational environment. Karl Weick has a particularly developed approach which includes this idea and the processual perspective as part of an approach to organisational analysis (Weick 1979). By stressing that all organisations rest on personal interaction Weick diminishes the idea that organisations are somehow 'real' beyond patterns of personal interaction conditioned by the inability of people to process all of the information that they receive. This social psychological model succeeds in being a coherent and self-contained alternative to the orthodox views of how fully 'rational' organisations function that are so beloved of economists, amongst others. The key processes in this model are ecological change, enactment, selection and retention.

The individual copes with being 'swamped' by information by reacting only to changes in the flow of experience and activity. This is the process of enactment where change in the stream of experience is isolated for further attention. The process of selection involves the imposition of various cognitive structures in an effort to reduce the equivocality of the enacted events. These cognitive structures Weick calls 'cause maps', which are interrelated patterns of variables derived from past experience. Such structures may provide reasonable interpretations of the current puzzling events or they may create more confusion; in the first case the cognitive structures tend to be retained, in the second they may be eliminated. This last process is retention, which is the storage of the new cause maps, or 'enacted environments':

> An enacted environment is a punctuated and connected summary of a previously equivocal display. It is a sensible version of what the equivocality is about, although other versions could have been constructed

> (Weick 1979: 131)

Within a group where a set of enacted events has to be given meaning there are likely to be several ways of interpreting such events, given the variety among individual cause maps. Negotiation within the group or organisation will take place which reduces the equivocality of meaning of the events and enables a move towards a common understanding – a common cause map for the interpretation of the enacted events.

This process of internal negotiation progressively shifts the members of an organisation towards a similarity of outlook, and I would suggest that this similarity is the basis for saying an organisation has a 'culture'. Culture is simply a pattern in the views and behaviour of a specific group of people and it should follow that a culture will rarely be a unique and well-defined entity, even to those inside the organisation, since it is coherent only as far as internal negotiation has reduced the equivocality of meaning of enacted events – a process which may never complete and which includes individuals for different periods of time.

In a sense Weick's model is a development of the commonsense finding that sensory perception cannot be separated from mental preconceptions – belief mediates perception and so there is no such thing as 'pure' observation distinct from the belief structure of the observer. Yet the set of beliefs that mediate perception are themselves selectively changed by the process of perception, and so there is this process where the model which modifies perception is itself updated – in Weick's terms an updating of the 'cause map'.

If this model of individual and organisational behaviour is accepted then meaningful environments are outputs of organising rather than inputs to it. The firm–environment boundary has no 'real' existence outside the nature and patterns of interaction between individuals, and individuals in the firm may experience large parts of their own firm as their 'environment', about which they have beliefs but no personal interactions, while their personal network may extend into other firms. It is useful to stress that this is in contrast to the usual view that the firm–environment boundary is a very real separation (one thinks of Chandler 1962 and 1976 with his senior managers rationally planning firm response to unequivocal environmental changes). In this study many of those interviewed had personal interactions outside of their own firms which they saw as important parts of their jobs, whereas they would not know the majority of the individuals who comprised their 'own' organisation.

Another feature of this model is that if we want to understand management action, we do not have to concern ourselves much about whether beliefs of managers are 'true' outside of the context of the organisation and its actions. We will wish to trace their own beliefs

– their cause maps – in a way that renders the process of managing understandable in their own terms and to others. The heuristics which were referred to earlier are an active expression of these cause maps drawn from previous experience.

The reason for setting out this social psychological model in some detail is that I do not believe it possible to examine innovation without examining how people think and create in an organisational context. To examine the process of innovation we need to examine the relevant parts of the cause maps of the managers who were closely involved in the novel food projects. They themselves will have selected the events that they believed were major influences on the project and there is a need to understand their actions as they themselves understand them. This is especially the case since, as the next section argues, there is no 'theory' of innovation, rather a complex of factors that influence the process of innovating.

INNOVATION CASE STUDIES

This book focuses on a specific type of innovation and so is a type of case study. Since it is sometimes mistakenly argued that case studies are only of value to those interested in the particular case, it is worthwhile setting out what value we can expect from a case study into innovation. The term 'case study' covers a vast range of research forms, so that it can be defined as 'an umbrella term for a family of research methods having in common the decision to focus an enquiry around an instance' (Adelman *et al.* 1977: 140). So if research is described as a case study, this does not imply the use of a standard research method or specify the arrangement of the content of the study; both the method employed and the findings of the case study are influenced by the assumptions and beliefs of the researcher. This understanding is reinforced when the scale of the reduction in information content is considered. It has been observed that a typical study might be read in a period a thousand times shorter than the time span of the original series of events (Burke 1970), so the selection as well as the presentation of events will depend, to an extent, on the concerns of the writer. So it makes sense to be as explicit as possible about the key assumptions and criteria used in the compilation of this study of innovation.

If there were fixed theories on innovation, then a case study could be justified purely as an attempt to disprove or verify a theory in that incidence. As is shown later, if we exclude the simplest, there are no theories, rather a common conclusion of others is that the process of innovation is complex and wrapped up in the nature of the people and organisations involved in it.

The more complex the questions and issues at the start of a case study, the greater the scope for complexity in its analysis and use. This variation in the complexity of case studies is one of their most interesting features; metaphors and anecdotes can be seen as being at one end of a range of complexity. The more intricate case studies will allow a number of interpretations of the events they describe, although they may then be eclectic about the conclusions they draw. In such a case study there will be a group of perhaps poorly connected but independently valid interests and expectations of relationships which do not add up to something with the status of a 'theory'. In case studies of this type the problem becomes how to cope with the many possible relationships within the material of the case study.

It is this type that is favoured in this thesis, and for the following reason. If what Gold refers to as the 'synoptic model' (Gold 1981) is rejected as a poor representation of the innovation process, it is because innovation is a complex process, although it does involve, as the synoptic model implies, some matching of technical and market opportunities. The requirement is not for more research which strains to constrain empirical evidence into the narrow waist-coat of available models. It is rather for the systematic sorting and display of the factors that affect the innovation process and the statement of likely connections between them in specific cases. That is, there is a need for work which can help a move towards better understanding and perhaps even a model of innovation as a complex process.

It should be said that there are various methodological 'evils' associated with the case study, and if these were to be listed they would certainly include the supposed inherent 'unrepresentativeness' of case studies and the use of inductive reasoning – the practice of making generalisations based on a single incidence. An articulation of these beliefs can be found in Popper (1972) or Medawar (1979).

Following Kuhn, most scientists are involved in 'normal science'; the investigation of cases in the attempt to disprove theories that are long-lived compared to those in the social sciences (Kuhn 1970). This predisposes them to be hostile to the use of case study to contribute towards theory or models that have wider application than the case study itself. Extrapolation from the case study is 'based on the validity of the analysis rather than the representativeness of the events' (Mitchell 1983).

More simply, arguments will be judged by common sense and their inherent likeliness. An intriguing example is given by Clyde-Mitchell, in the story of the psychologist who found a statistically significant correlation between stomach disorders and the tendency to perceive frog-shapes in Rorschach inkblots (Mitchell 1983). The psychologist's Freudian explanation was that the frog was anal-symbolic and demonstrated that the stomach disorders were linked to anal-obsessive personalities. This analysis was seen as inherently unlikely by his clinical psychologist colleagues and no follow up research was carried out, despite the impeccable statistics and the 'scientific' use of Freudian theory. In this case it was the validity of the analysis which was at fault, not the research method – even with an 'impeccable' methodological approach, ridiculous conclusions can be reached. Researchers can become obsessed with so-called 'scientific method' to the detriment of common sense. The last word on the legitimacy of generalising from cases belongs to a veteran user of the method in education research.

> It might be asked whether the generalisations produced by case study are stronger or weaker than those of experimental research. Stronger or weaker they tend to be different . . . in practice the most important differences are in the way claims are made against truth and in the demands made upon the reader. Experimental research 'guarantees' the veracity of its generalisations by reference to formal theories and hands them on to the reader; case study research offers a surrogate experience and invites the reader to underwrite the account, by appealing to his tacit knowledge of the human situation.
>
> (Adelman *et al.* 1977: 143).

So far two key points have been made in relation to innovation case studies:

1 The number of factors considered in the study and their inter-relationships should reflect assumptions about likely influences on the process of innovation rather than be determined by a single model or theory.
2 It will be valid to make informed comment about the process of innovation in general, based on the case study.

This brings us to other studies of innovation and the lessons that they provide. An especially useful review of innovation diffusion studies by Gold serves as a basis for setting out our assumptions about the innovation process, and these are summarised as follows (abstracted from Gold 1981):

1 Innovation is rarely discrete and does not develop in isolation from other innovations and products – it has a 'technological context'.
2 Innovation often arises because of other, larger aims, such as an attempt to produce a novel product or process.
3 The innovation and its environment are not static during the period of development and subsequent diffusion in the environment, as many models implicitly assume.
4 There are few studies of failed innovation and therefore little chance of an informed debate on the reasons for innovation success and failure.
5 There are few studies which compare reasons for the adoption of the same innovation in different companies.
6 There are few studies of the early part of the innovation generation process and of the reasons why managers and scientists pick one innovation over another for development.

There is a particular widespread assumption about the innovation process, which Gold calls the 'synoptic model' of innovation. In this model the R & D department is the site where basic science is 'applied' to produce new technology; this technology is then matched with market needs to produce innovation. This simple, linear model is not 'wrong', but is guilty of gross oversimplification of the complexity of the innovation process. Science certainly cannot be assumed to have precedence over technology, while the idea of a market may well play a part in the innovation process from its very beginning. The structure and organisation of the firm may also influence project decisions, as Margaret Graham concludes after her detailed discussion of the reasons for RCA's development of the Videodisc,

> the choice of technical approach to a given innovation often relies more on the internal needs and preferences of various parts of the corporation than on a sense of a need in the market place. Jobs, key skills, use of readily available equipment, shared characteristics with other projects and fulfilment of individual organisational goals are all legitimate internal needs that can influence choices made about a technology as much as, or more than information about the market.
>
> (Graham 1986: 3)

In another example of the complexity of the innovation process, Rosenbloom and Abernathy have attributed the decline of the US consumer electronics industry to management beliefs and consequent behaviour (Rosenbloom and Abernathy 1982). US management believed that consumer electronics was a 'mature' industry and so they relied on advertising, rather than technical excellence and continued innovation, to sell VCR products. In contrast, the Japanese were not deterred by market rejection of their innovative VCR prototypes and they concentrated on improved product quality through production process innovation. The result was that the Japanese effectively captured the US market for VCR products, as they had done for other consumer electronics items. The contrasting management belief structures led to different treatments of technology and hence different competitive results.

In a fascinating British study, by Langrish and his co-workers, every firm that had won the Queen's Award for Industry over a period of two years was investigated and the attempt made to understand why their innovations had been successful (Langrish *et al.* 1972). In their conclusions the complexity of factors that affect innovation were stressed and the comment made that 'new productive process is the historical outcome of many strands of events' (Langrish *et al.* 1972: 7).

If innovation is examined from the point of view of the R & D or project manager, certain problems with innovative projects occur repeatedly. Mansfield, a writer with much experience of research into R & D and project management, has produced the following list of research 'commandments' (Mansfield 1977: 6):

1 Research should be related to a company's goals.
2 Relate research to competitors, customers and the scientific community.
3 Do not underestimate the uncertainty of projects.
4 Ensure proper transfer of new technology to operations.
5 Do not allow research to become isolated from development.
6 Do not allow degeneration of R & D to a 'technical servicing department'.
7 Do not focus solely on R & D – use innovation in other firms and industries and avoid the 'not invented here' syndrome.

At first glance this list appears to be a statement of the obvious, but given that it is a distillation of Mansfield's experience of how projects go wrong, it is perhaps better to treat it as a warning of how difficult it is to achieve the 'obvious' in practice. So it further reinforces the idea that the effort of trying to understand how managers and firms organise the innovation process is a worthwhile one.

We have moved from macro-economic models of technical change to technical change within the firm and manager decision-making. If they have a form of bounded rationality as outlined previously, we still want to elaborate further how they deal with the multiple influences on the innovation process. The point has been made by Bessant and Grunt that a full list of influences is not really possible because in each case of innovation, managers will ascribe causative power to a different range of factors (Bessant and Grunt 1985). The range of factors is wide, however, with Bessant and Grunt having their own list of influences on the innovation process culled from their own and other case studies. They range from those concerned with personality, such as values, beliefs, experience and education, those concerned with the local environment, such as operating division and firm, and those external to the firm.

Braun suggests what he calls 'constellation theory' in an attempt to partially structure how multiple factors affect innovation (Braun 1981). Decisions by managers to proceed with innovation development only occur when they perceive patterns among the relevant influencing factors. So innovation writers have moved towards a characterisation of the process of innovation in similar terms to those used by Weick when analysing the role of individuals in organisations. The important feature here is that managers enact, select and retain experience which they believe affects the chances of innovation success – and for this reason in this study there is a focus on the managers' beliefs and explanations pertaining to the innovation process.

What has been set out in this chapter is a loose conceptual framework for the study of the innovation process which makes no prior assertions about causality in innovation, but retains the awareness that of the many possible influences on innovation, it is those to which managers ascribe importance that are of interest here.

THE RESEARCH

The research followed a developing 'network' of manager contacts, with one contact suggesting the names of several other managers who might cover, for example, the market or technical sides of the project in other companies. These networks tended to 'close' as later interviewees referred back to earlier ones. The origin of the manager network was their shared experience as microbiologists at university and the personal acquaintances they made on the fermentation conference circuit. This was sufficient for them to know the names and positions of most of their counterparts in other organisations. As Foot-Whyte has noted, informants at 'connections' in the network are often especially valuable and SugarCo were particularly fruitful in providing contacts in other companies (Foot-Whyte 1960).

Interviews were 'semi-structured' – they sought to allow the managers to tell the story as they saw it, but there were simple questions on what had happened and in what order, which built a chronology or 'event schema' (Pettigrew 1979) for the projects. Later interviews also included feedback in the form of key questions which drew on earlier interviews and the managers' categories of analysis of the projects. This approach follows from the argument set out earlier; if individuals retain selected experience in the form of cause maps, interviews involve an effort to make those maps articulate. What one can expect from interviews is the managers' selection of key events and new knowledge, influenced by their previous experience. The approach is strongly supported by Hickson *et al.*, who compare an ethnographic study of a social setting with an attempt to understand the setting through selective semi-structured interviews. They comment that

> It was found that the essentials of problems, interests and processes could be gathered by interview. Interviews give an outline narrative of main events . . . through the answers to a series of questions about what happened, without it being necessary to discover every incident. The hindsight story that is forthcoming . . . is the same in main events and characteristics, just less cluttered with detail

(Hickson *et al.* 1986: 25)

Comparative analysis (Glaser and Strauss 1967) of the transcribed interviews was chosen as a method which lent rigour to the process of sorting material and which aided the development of additional insights. This method involves the development of categories for sections of the text by the systematic comparison of all text content. A re-sorting of the text according to categories yields an analytical structure which in this case aided the generation of the chapter structure and which broadly reflects the concerns of the managers as they relate to the process of innovation.

According to Glaser and Strauss the result of comparative analysis of such data is 'grounded theory'; comparative analysis is productive of insights and conclusions rooted or 'grounded' in the empirical material. These may form the basis of an argument that the case study has relevance outside of itself. In this case it will be considered valid to make comment on the general process of innovation informed by the detailed analysis of the case study.

2 Introduction to the technology

THE CASE OF SINGLE-CELL PROTEIN INNOVATION

The innovations on which this book is based are those by which companies attempted to produce a radical change in food and feed production. They attempted to substitute a form of industrial fermentation technology for agriculture as a method of producing protein. According to the innovation taxonomy they can be called 'attempted radical innovations', which, had they been successful on any scale, would have had a major economic impact on the animal feed business.

The technology was chosen for its interest on several levels. It is a study of both failure and success within the same technological field. Companies from very different industries entered the area, including firms from the food, oil and chemical industries. A variety of stages in the innovation process were reached in different companies, from laboratory research to full-scale commercial plant and the marketing of the product. Market conceptions changed through the development of projects and differed between projects. All the projects impinged on firm strategic concerns, but in different ways at different times. The technology itself differed at a detailed level between firms.

The main geographical boundary to this study is Western Europe, where the major commercial developments took place, although reference is made to developments elsewhere, especially in Japan and the Soviet Union, because these did have an impact on the management of the European projects.

Within Western Europe the firms were selected for study if they had attempted to use fermentation technology to produce new foods or feeds. Within this group there was huge variation in the manner in which the technology was exploited, so much so that managers would sometimes regard other firms in the group as in another business. Paradoxically, this variety of approaches to the technology is one of the strengths of the study. The variety suggests something about the nature of technology itself; it is given a structure, a detailed form by people and their organisations. It is the motivation and the perception of interests of these people in the innovating firms that will be investigated in this study.

The following chapters focus on the purpose-built single-cell protein (SCP) plant, built because it was believed there was a large and growing first world market for SCP. This excludes two versions of the technology that were developed to produce SCP, from carbohydrate-rich wastes and for the Third World – these are described in the Appendix and show the adaptability of the technology.

In this chapter the history of the SCP projects is summarised and the technology and relevant science is described in some detail. The chronological overview and description of the production process principles is useful when managers come to talk about how decisions were made in each step towards the commercialisation of their products. For reference, there are tables of the companies that undertook projects, the size of plant that they developed and the names of their different products.

A SHORT HISTORY OF SINGLE CELL PROTEIN[1]

The name single-cell protein, or SCP, was given to the industrial processes for fermenting micro-organisms by Nevin Scrimshaw, an academic at MIT in the late 1960s. This was the golden age of SCP, when many oil companies were starting research on the feasibility of processes based on yeast, bacteria or fungi.[2] The aim was to produce high-protein foods or feeds through the industrial fermentation of these micro-organisms on some kind of cheap substrate, usually a hydrocarbon.

Scientific papers were published as long ago as the late nineteenth century on the ability of certain bacteria to grow on methanol. During the Second World War both the Soviet Union and the Third Reich had hurriedly constructed food yeast manufacturing plants to provide protein supplements to the diets of their industrial populations. Germany had first attempted to construct such plants in the First World War, based on the fermentation of Candida Utilis yeast. These plants only reached half the German target production figures because of technical problems (Solomons 1983).

During the defence of Stalingrad Soviet soldiers ate dogs, cats, rats . . . and yeast protein food supplements. One of the managers even suggested that the Soviet manufacture of yeast foods should be added to the list of factors that 'could have tipped the balance' during the Second World War. It is from this time that the Soviet Union was interested in single-cell proteins, and unlike the West, the Soviet Union continued to operate more plants than any other country in the world.

At the end of the war the capture of German SCP-manufacturing technology was a priority interest for both the Soviets and the Western powers, given its strategic importance in feeding the German civilian population. Immediately after the war and in an international climate of food shortage, SCP plants continued to be built to produce human food supplements. The first commercial processes where single-cell organisms were deliberately grown and harvested in quantity were installed in the 1920s and used sulphite liquor, which is a waste product from paper pulp manufacturing. These supplied yeast biomass as a by-product which was sold as an animal-feed additive. In 1948 Candida yeast was grown on waste sulphite pulp liquor in the United States using German-type technology; later a second plant was built of 5,000 tonnes per annum (t/a) capacity, which has only recently ceased operation. In 1947 the British government financed a project to grow Candida Utilis on sugar-cane molasses in the West Indies as a protein-vitamin supplement for the indigenous population. The project eventually failed, apparently due to technical problems and undercapitalization (Solomons 1983).

But once the food shortages of the war and the immediate post-war years had passed, interest dropped away in yeast as a food. It would only revive again when world famine and food shortages once again appeared imminent, and this time the bulk of the projects would involve the fermentation of single-celled organisms on hydro-carbon substrates. The trigger for the increase in interest came when a French research scientist called Champagnat, who was contracted to work for BP, announced his team's successful cultivation of yeasts on petroleum. At the time it was known that micro-organisms could grow on water-insoluble substrates such as petroleum, but there was an assumption among scientists that these organisms would be rather specialised and would grow only slowly and so would be of little industrial interest (Solomons 1983). The BP announcement stimulated world-wide interest in these organisms, particularly among the oil companies.

During the 1960s oil was cheap and protein, in the form of

soya-meal, was relatively expensive. More important, the trend in the oil price was downwards, that of soya upwards or expected to be upwards. Chemical companies with access to cheap hydrocarbon substrates began research programmes into SCP; ICI and Norsk Hydro took North Sea gas as their starting point for bioprotein manufacture. The expectation of an increasing world protein 'gap' between production and requirements was a further reason for developing the technology. In the late 1960s to early 1970s world population and therefore demand for food was expected to grow faster than conventional agriculture could expand its output and it was thought that there would be a shortage of protein in many Third World diets. These forecasts of world-wide protein shortages seemed to guarantee the economic success of new protein manufacturing processes. It was this expectation of a food shortage that encouraged some of the food companies into novel protein manufacture. The food companies used non-oil derived substrates such as starch and glucose, where these were being produced as surplus to other food-manufacturing processes. The fermentation of the substrates would enrich the protein content of these by-products and offer the prospect of turning them into complete foods with higher value.

Another group of companies continued to manufacture SCP as a means to solve waste disposal problems linked to their principal activities. This group includes wood pulp, confectionery, dairy and palm oil producers, all of whom produce large volumes of carbohydrate-rich waste water. Most were uninterested in the SCP conference circuit, preferring to buy in technology from equipment manufacturing firms, and it was the latter that tended to do the research, build the pilot plants and attend conferences on SCP. The largest capacity waste processing plant would be less than 10,000 t/a and all of them were limited by the volume of waste produced. This compared with target capacities for the single-purpose SCP plants of around 100,000 t/a.

The oil price rise shocks of 1973 and 1979 overturned the economics of the hydrocarbon-based SCP processes and during the period 1973–82 research and development was abandoned by almost all the oil and chemical companies. The soya price fell in real terms where it had been expected to rise during the 1970s, and this seemed to finish hopes of an economically viable process. However, with the fall in hydrocarbon and energy prices in 1986 the economics have become more favourable. There is one project run by Dansk Bioprotein in Denmark which aimed to produce SCP for animal feed from North Sea gas by late 1991. The food and waste-based processes were not severely affected by the change in oil price, but the cheapening of the price of soya in the 1970s and the development of new waste treatment technologies undermined the economics of some waste processes.

There is only one novel food product now on the market and this is 'Quorn', which is a mycelium fungus grown on a glucose substrate by Marlow Foods, a company which was jointly owned by RHM and ICI and is now completely ICI-owned. By 1989 there were over forty prepared dishes containing Quorn which were being sold through various supermarkets and chain stores in the UK (Spencer 1989). By this time it was the intention of RHM and ICI that Quorn would become a major, internationally traded foodstuff. Another step towards this goal was taken in 1993 when ICI Zeneca (the company which now comprises the pharmaceuticals and biotechnology activities of ICI) announced a £26.5 million investment in a new plant at Belasis, Cleveland, to produce an extra 7,000 tonnes of Quorn per annum.

The success of any of these projects would mean the addition of a new food to the human food chain. Either animals would eat the novel proteins and people would eat the animals, or else people would eat the proteins directly. The success (albeit on a small scale at present) of Quorn is the first time since the potato that a Western population has begun to eat an entirely novel source of food, that is, one not previously eaten by any other human population.

The attempt to commercialise single-cell protein can be seen as the first development in what would now be called biotechnology, and many of the scientists who originally worked on the SCP projects are now to be found in other biotechnology activities. Some of the technology developed by the SCP projects has been applied to other projects in the bio-industrial field which are still current, but it is likely that much of the technology will never be used because it is too specific to a certain type of biological production; continuous fermentation used to produce a low-priced, bulk commodity product. At present, low-priced, biological commodities of all kinds are produced more cheaply through agriculture than industrial fermentation.

Between the time BP announced its interest until the present time, roughly £1 billion in today's money has been spent on SCP projects in the West. The three companies that built large commercial plants incurred most of this development cost; BP, ICI and Liquichimica. There have also been hundreds of small research groups all over the world, mostly in universities, who have investigated a wide range of possible organism and substrate combinations. These substrates include peat, all kinds of vegetable processing wastes, animal waste slurries, oxidised polyethylene and fish oil. Solomons (1983) lists twenty-seven references to unusual substrate-organism SCP research, which has now almost entirely faded away.

COMPANIES

Table 2.1 shows which companies were involved in the technology and the stage to which they pursued development. Although there was variation in the technological complexity of the pilot plants and full-scale plants, the table ignores this and serves as a rough guide as to how far each company progressed. It can be used to estimate the expenditure of the companies and the relative importance of the projects. The cost of a pilot was of the order of millions of pounds – that for a full scale, 100,000 t/a plant was between £200 million and £300 million, if all the development work including pilot plant construction is included in the estimate. Only three companies spent this amount on SCP: ICI, the BP–ANIC combination and the Italian company Liquichimica. At the opposite end of this scale of expenditure was SugarCo and the other waste substrate-based processes, with expenditures of a few million pounds on low-capital-intensity technology.

For any of these projects research involved growing experimental

Table 2.1 Company, product and approximate size of plant built

Company	Product	Pilot plant	Full-scale plant
ICI	Pruteen	1,000	70,000
BP	Toprina	4,000	100,000
Liquichimica	Liquipron	Yes	100,000
Hoechst	Probion	Yes	
Shell		Yes	
SugarCo		Yes	
Esso–Nestlé		Yes	
Pekilo		Yes	
RHM	Quorn	Yes	
Norsk-Hydro		Yes	
Unilever			
Amoco	Torutein	7,500	
Philips Petr.	Provosteen	30,000	
Swedish Sugar	Symba	10,000	
Cell. Attishlz		Yes	
Bel Industrie	Protibel	Yes	
IFP/ELF/CFG		Yes	
Romania/Japan		Yes	

cultures of micro-organisms in batches on agar jelly or in shake flasks, but this yielded no sizeable quantities of product. The nutritional and toxicological research that was essential for the development of these products required more sizeable quantities that could be produced from a pilot plant, or experimental fermenter, of around 1,000 t/a capacity. A pilot plant enabled the training of technicians and aided the scaling-up of fermentation to a fully commercial plant – Liquichimica built a pilot plant with the sole purpose of training the technicians who would have to operate their full-scale plant. The jump in scale from a pilot plant to a full-scale plant of between 50,000 and 100,000 t/a capacity still assumes that the larger-scale fermentation will pose no radical engineering problems – the more cautious companies built semi-industrial plant to test the viability of full-scale production, others jumped to full-scale from pilot plant. Product from the BP semi-industrial plant was sold on a continuous basis on the open market, capacity for these plants was in the 10,000 to 20,000 t/a range.

The plant constructed for Cellulose Attisholz and the Pekilo plants were 'semi-industrial'. These had a different economic justification from the hydrocarbon-based projects, and a semi-industrial size plant was as large as a waste-based protein plant could hope to become, given its dependence on the quantity of waste or by-product produced by the primary process.

SCIENCE BASE: MICRO-ORGANISMS

The two types of single-celled organisms used to grow protein were yeast and bacteria. Bacteria are of the order of 1–2 micrometres, or millionths of a metre, and reproduce by binary cell division with a cell division time of 1–2 hours. Yeast cells are larger, have thicker cell walls, a lower protein content and slower growth rates than bacteria. Their cell size is in the range of from 5–10 micrometres and they reproduce by 'budding' smaller cells from the parent cell membrane. Yeast are, in effect, species of single cell fungi. Fungi are multicellular and have a thread-like, filamentous structure. They grow more slowly than yeast and bacteria and have even lower protein content.

The faster any organism grows, the greater the percentage of ribonucleic acid (RNA) in the cell mass. RNA is the nucleic protein of the cell which transmits genetic information from DNA as part of the process of creating new proteins. If the cells are growing and dividing rapidly, a greater proportion of total cell mass is given over to protein production in the form of RNA; RNA and DNA have no nutritional value to mammals and are simply broken down by the liver to uric acid and excreted. Worse, if a human diet contains too high a percentage of nucleic acids, the uric acid burden on the kidneys may be too great and uric acid may deposit out of the urine as small stones or crystals in the kidneys and joints – the cause of gout.

In a particular species of yeast or bacteria there are many strains. So Candida Lippolytica, Candida Albicans and Candida Tropicalis are all yeast species, but each contains many strains, each with its own optimal growing conditions of acidity (pH), temperature, pressure, nutrient requirements and product concentrations. Even within the same species, different strains metabolise different products, so different strains may be toxic or non-toxic to animals and human beings. These strains may not even be named in the micro-biological texts, if there has been no reason for microbiologists to take an interest in them. This was the case for ICI's bacterium, which they named Methylophilus Methylotrophus, a member of the Pseudomonas genus.

A typical equation for cell growth is the following for a methanol solution:

$$CH_3OH + NH_3 + \text{nutrients} + O_2 = \text{Cells} + CO_2 + \text{extra-cellular products}$$

The biochemical pathways of these bacteria are adapted to the metabolism of methanol to provide them with energy for growth, and they synthesise their proteins out of the ammonia. While the carbon dioxide bubbles out of solution, the extra-cellular products excreted by the cells stay in the methanol solution. They may contain growth-limiting substances and they also represent a loss of carbon from the cell mass. Differing growth conditions will change the cells' metabolism and they can be manipulated to minimise the production of these products and maximise cell weight-gain.

GROWTH SUBSTRATES AND HOW TO CHOOSE ONE

The substrate is the source of carbon and energy which the micro-organism will use to build its cells and metabolise proteins. Any substrate will require nutritive elements to be added to aid growth, in the form of nitrogen, phosphorus, trace elements and so on. These nutritive elements were supplied as aqueous solutions, so the substrate needed to be water-soluble or, in the case of the n-alkanes, prepared as an emulsion. Nitrogen was generally supplied as aqueous

Table 2.2 Substrate, organism and company

Substrate	Micro-organism	Companies
Methanol	Bacterium	ICI, Shell
	Yeast	Elf-Industrie Français du Pétrole
Methane (North Sea gas)	Bacterial culture	Shell, Dansk Bioprotein
N-alkanes	Yeast	BP
		Italproteine
		USSR
		Liquichimica
		Romania
	Bacterium	Chinese Petroleum Corporation (Taiwan)
Gas-oil	Yeast	Elf-Institut Français du Pétrole
		Indian Institute of Petroleum
		BP France
Carbohydrate	Fungus	Ranks Hovis McDougall (RHM)
		SugarCo
	Yeast	Swedish Sugar Corporation
Lactose	Yeast	Bel Industrie (dairies)

ammonia and sulphur and phosphorus as sulphuric and phosphoric acid.

All the companies were looking for a substrate which was available in large quantities, was cheap and capable of being produced in a very pure form. Other factors affecting the choice of substrate were price, availability and control over supply. In Table 2.2 the substrate and organism of each company are listed. It is immediately apparent that the business of the company coincides with the type of substrate they chose. For example, ICI was able to control large supplies of natural gas, so they chose methanol as a substrate, which they synthesised from natural gas. The principal substrates are described in Box 2.1.

CONSIDERATIONS IN THE CHOICE OF ORGANISM

Many hundreds of different types of micro-organism were investigated by ICI, BP and RHM and samples of each were grown in shake-flasks and on agar plates containing nutritive media and substrate. To start with, many strains of each organism were present on the plates, but after several rounds of being cultured the strains most adapted to the environmental conditions imposed on them would be left. In this way a short list of candidate organisms, all with desirable features, could

Box 2.1 The principal substrates

Methane	The simplest hydrocarbon, a gas at room temperature, barely soluble in water, a major constituent of natural gas, occurs naturally with oil reserves.
Methanol	The simplest alcohol (methylated spirit), a liquid at room temperature. Soluble in water, which is its big advantage over methane. Toxic.
N-alkanes or normal alkanes	Alkanes are the simplest group of hydrocarbon compounds, of which methane, ethane, propane and butane are members. Alkanes are obtained from crude oil, which can vary in its alkane content. The 'normal' signifies that these are straight-chain molecules, with a length of carbon atoms being surrounded by hydrogen atoms. Different alkanes are defined by the number of carbon atoms in the chain, so the notation C_7–C_{20} includes all those alkanes having between seven and twenty carbon atoms in the chain. The importance of the straight chain alkanes is that the microbes which degrade them use an enzyme which acts on the methyl (CH_3) end of the molecule. Carbon–carbon junctions in isomeric, branched forms of the molecules are 'indigestible'. Those used in protein fermentation are liquid at room temperature and must form an emulsion before they can be fermented; these are in the range C_5–C_{24}, with the heavier alkanes being soluble in the lighter fractions.
Gas-oil	This is the product of the first distillation of crude oil. It is a complex mix of hydrocarbons, some aromatic (benzene-ring based), alkanes and others. It differs from crude oil in that the most volatile fractions and the solid (bituminous) parts have been removed. As a result of the low level of refining, it is cheaper than the n-alkanes, but known carcinogens feature among its impurities.
Carbohydrate	These are carbon–hydrogen–oxygen compounds of lower calorific value than the hydrocarbons. Plant processing wastes can be used, such as the sucrose in molasses or the lignin-related wastes in spent sulphite liquor from wood-pulp processing. Food grade materials can be used; for example, cereal starch can be hydrolysed to water-soluble glucose, which is then fermented.

be obtained. Carbon conversion efficiency is one such desirable property as is the proportion of carbon in the substrate that is converted into carbon in the cell mass. By making the substrate the growth-limiting factor in the cultures, strains which were efficient carbon converters would be 'unnaturally' selected so that they would eventually dominate the culture, perhaps as pure strains. In this way, the carbon conversion efficiency of the ICI organisms was raised from 45 per cent to 60 per cent. Once the optimum growing conditions were determined a pilot plant fermenter could be built to test out the short-listed organisms in a continuous growth process, rather than in the batch production of plates and shake flasks.

The choice of a strain of micro-organism was governed by a much larger number of factors than the choice of substrate. The first choice was between a yeast and a bacterium or fungus and the principal differences between these three types of micro-organism are outlined below – many of these differences have implications for the design of production plant that could be built.

1 *Speed of growth.* Bacteria grow most quickly, then yeast, then fungi. The more rapid the growth of the organism the higher the output of a plant with fixed fermenter capacity. So a fast-growing organism offers the chance of a higher rate of return on a fixed-capital investment.

2 *Protein content.* Another advantage of bacteria is the higher protein content. Table 2.3 below shows the percentage of crude protein that each type of micro-organism contains. This advantage is offset by the faster-growing bacteria also having a higher percentage of ribonucleic acid in their cells, which has no nutritive value. If the protein is to be used for human consumption, this may need to be partially removed. The rest of the cell mass is made up of lipids (fats) and carbohydrate.

3 *Size of organism.* Bacteria, due to their small size, were difficult to centrifuge and presented a harvesting problem. ICI devised a technique to break open the cells and cause the contents to clump together into flocs of coagulated protein. These would be large enough to centrifuge out of the protein–water suspension, although this treatment added to processing costs and affected the nature of the proteins. On the other hand yeasts could be centrifuged out of their growth medium in perhaps two steps, to form a creamy-white slurry. Fungi were the easiest of all to harvest because of their large filaments, which could be collected by one simple filtration step.

4 *Resistance to contamination.* Some organisms would be more

prone to infection with foreign micro-organisms which would feed off either the extra-cellular products or the dead cells of the principal organism. The latter was the case with ICI's methanol-consuming bacterium. This sensitivity to infection may require sterility in the fermentation process, and sterility requires a whole new range of engineering standards and products, such as sterile valves and sterile welds and seals. Once the decision to have a sterile process has been taken, it is an influence on fermenter design. The major fermenter types are the stir-tank and the air-lift fermenter, the latter being a type of tower fermenter. Air-lift fermenters, with no moving mechanical parts, are easier to keep sterile than the stir-tank variety.

5 *Carbon-conversion efficiency*. This has already been described as an important efficiency consideration. Another result of a poor conversion efficiency is that the heat of fermentation will be greater, as the carbon in the substrate is completely oxidised to carbon dioxide. This heat must be removed if the optimum growing temperature is to be maintained, and so this characteristic of the micro-organism affects the design, cost and capacity of the cooling system.

6 *Consumer acceptability*. The final product must win the approval of customers, whether animal feed suppliers or the eating public. The companies knew that common perceptions of each type of organism would be used to judge their products. So a food derived from a bacterium might be suspect because bacteria cause disease and decay – the popular image of bacteria is 'ugh – germs'. Yeast are already eaten directly in the form of yeast extract by human beings and waste-substrate-derived yeast slurries have been eaten by animals ever since brewing became an industry. So yeast should be more acceptable if the company were to venture into human food manufacturing. Top of the image range are fungi, at least compared to yeast and bacteria. Fungi are widely accepted as part of many foods, such as cheese. If the name is a problem, then 'mushrooms' are fungi, and this name could plausibly be used to describe any fungal food.

Table 2.3 Percentage of protein in bacteria, yeast and fungi

Organism	% Protein
Bacteria	80
Yeast	60
Fungi	45

Source: ICI 1974: 8

NOVEL PROTEIN PRODUCTION TECHNOLOGY[3]

There were three key stages to the production process.

1 Medium preparation and sterilisation.
2 Fermentation.
3 Harvesting.

These can best be described by the example of ICI's process. In the ICI Pruteen process, the first stage was the preparation of an aqueous solution of methanol and nutrients. Mineral salts and acids were dissolved in the solution to supply the organism with nitrogen, phosphorus, sodium, potassium, magnesium and iron.

ICI had chosen to run a sterile fermentation process, so each of the fermenter inputs had to be heat-treated to sterility.

The ICI fermenter remains the most technically complex fermenter ever designed, an example of an air-lift fermenter that they called the 'pressure cycle' fermenter, a design which uses air under pressure to mix the fermenter contents and supply sufficient oxygen to the micro-organisms. An alternative was a stir tank fermenter of the sort that BP used, which because it uses a mechanical paddle to stir the fermenter contents is more difficult to keep sterile – foreign organisms can enter via the moving-part surfaces. The simplest fermenters were used in 'village technology' projects of the sort funded by SugarCo. These consisted of no more than a plastic or metal box with a lid.

In Figure 2.1 the arrows in the fermenter show the direction of circulation of the fluid. The blown-in air forms a low-density bubble–methanol solution mixture which rises in the outer sheath of the fermenter. The high hydrostatic pressure at the base of the fermenter tower allows oxygen to dissolve rapidly into solution from the many tiny air bubbles. As the bubble–solution mixture rises the pressure falls and waste carbon dioxide comes out of solution and into the tiny bubbles. In the gas disengagement section of the fermenter there is little hydrostatic pressure on the fluid, therefore the solubility of the gases is low. The carbon dioxide and oxygen-depleted air leave the surface of the liquid column leaving a high density, bubble-free liquid. This then descends in the inner vessel of the fermenter, shown as a tube in Figure 2.1. In this 'downcomer' section there is a heat exchanger which removes the heat of fermentation. At the base of the downcomer the high-density liquid is aerated once again and the cycle repeats itself, the methanol solution circulating continually.

Harvesting the bacteria from solution involved a flocculation step.

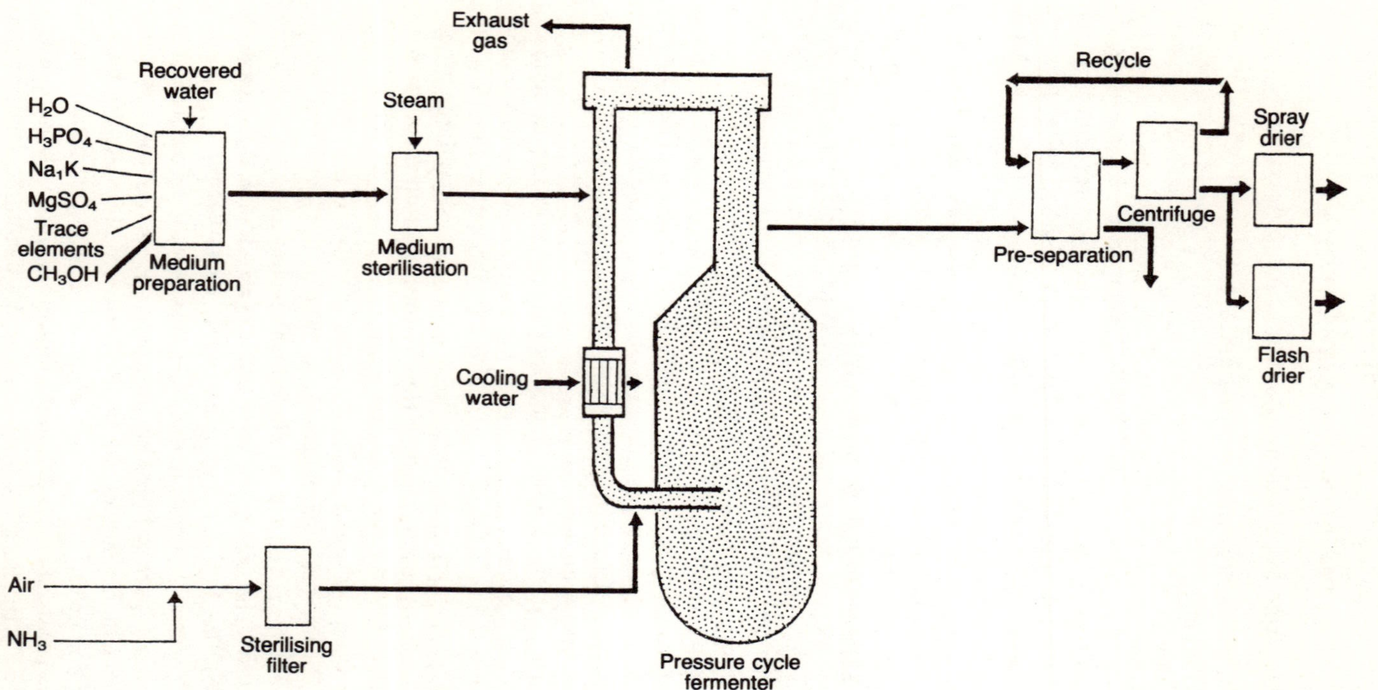

Figure 2.1 Flowsheet of Pruteen process
Source: ICI 1974: 10

This step used acid solutions and electric current to make the bacterial bodies clump together into larger flocs, which could then be further separated from the solution by several centrifuging steps, yielding a slurry which was 26 per cent bacterial protein. The last stage allowed the preparation of either a powdered or a granulated form of Pruteen. The powdered form was made by spray-drying; here the slurry was sprayed vertically into a chamber of heated air where the water evaporated to leave a dry powder. Flash-drying is similar but recycles partially dried slurry particles so that they clump together and form granules.

The final product is creamy coloured, odourless and tasteless. For the animal-feed market, this is of no consequence, but it is a result of the processing ICI used and their choice of a bacterium as organism. In effect the decision to market the protein as an animal feed is built in to the fabric of the plant – it was not designed to produce human-food-grade materials.

NUTRITION AND TOXICITY TESTING

Once an organism had been selected for its promising growth characteristics and sample cultures grown, the protein product was analysed for its nutritional and toxicological properties (Figure 2.2). An important guide to nutritional value was the amino acid profile of the protein (Table 2.4). This profile is a percentage breakdown of the essential amino acid content for every 100g of dry single-cell protein. Essential amino acids are the building-blocks of all other proteins, essential because animals cannot synthesise these amino

Table 2.4 Amino acid profile of common animal feed protein sources (in grams of amino acid/100grams of product – water content approx 10%)

Amino acid	Yeast protein	ICI protein	Fish-meal	Soya-meal
Isoleucine	2.8	3.1	3.2	2.2
Leucine	4.1	4.9	5.0	3.3
Phenylalanine	2.4	2.5	2.9	2.2
Threonine	2.7	3.3	3.0	1.9
Tryptophan	0.8	0.7	0.9	0.6
Tyrosine	2.0	2.2	2.3	1.6
Valine	3.3	3.9	3.7	2.3
Cystine	0.6	0.5	0.7	0.6
Methionine	1.0	1.8	1.9	0.6
Lysine	4.1	4.5	4.9	2.8

Source: ICI 1974: 16

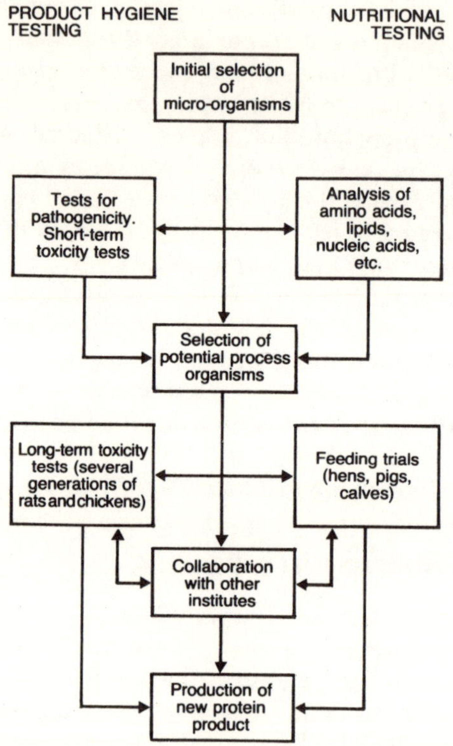

Figure 2.2 Summary chart of test procedures

acids for themselves. Instead they must either eat them or eat proteins which can be broken down into these amino acids in the digestive tract. The richer a protein is in these acids, the higher its value to the animal feed manufacturing companies.

Two of the essential amino acids contain sulphur and are the source of sulphur for all protein synthesis. Soya and other vegetable sources of protein tend to lack these two amino acids, lysine and methionine. SCP amino acid profiles show that SCP is relatively rich in lysine and methionine compared to the standard animal feed ingredient, soya-meal (Table 2.4), although not as rich as the standard high-protein component of animal feed, fish-meal. The relative percentage of the sought-after amino acids strongly affects the price of animal feeds, pushing the market price of SCP above that of soya but leaving it below fish-meal. Table 2.4 was used by ICI to argue that Pruteen was comparable to fish-meal in protein quality and so could command a fish-meal price, which is perhaps double that of soya-meal. Amino acid profiles also vary between strains of micro-organism and so the profile becomes another criterion for the selection of an organism.

Once the organism was being grown in sufficient quantity, nutritional trials began with animals, where the protein product was progressively substituted for the usual diet of soya- and fish-meal. Statistical comparisons were made with a control group in terms of daily live weight gain, and the food conversion of the animals. In the case of ICI's Pruteen, no significant differences in the two types of diet were observed, except for one set of experiments on pigs where the Pruteen diet produced a significantly improved performance in the final growing period.

Another major feature of the development of the novel food and feed proteins, toxicological testing, presented something of a problem to the various health authorities. Testing procedures and regulations are established for additives, but not for entirely new foods. The usual procedures for new additives or drugs involves feeding progressively larger numbers and varieties of experimental animals with doses hundreds of times larger than that expected to be administered to human beings, and if no toxic effects occur, the product moves on to human trials. The assumption is that if no ill effects occur in large doses, they are unlikely to occur in 'normal' doses, but a new food poses a testing problem because it cannot be administered to animals in multiple doses of this size.

ICI tested potential micro-organisms on mice, and if there was a toxic effect the organism was rejected. Then a second and more costly set of trials was undertaken where the protein product was mixed with other feeds and fed to chicks, rats and pigs and the results were compared with a control group of animals fed only on ordinary feeds. These tests were repeated over three generations of animals, to control for possible genetic defects and long-term deleterious effects. ICI believes the results showed Pruteen to be completely safe, and these tests have been accepted by the UK government as proof that Pruteen is not toxic to animals.

THE ECONOMICS OF MICROBIAL PROTEIN PRODUCTION

Producing fodder yeast from wastes has the advantage that the waste has either no value or a negative value, but a plant which is dedicated to making SCP must buy the substrate at world prices.

The key indices for the economics of the hydrocarbon-based SCP processes are the price of the substrate and the retail price of the product. The retail price must fall between the price of soya-meal and the price of fish-meal to stand a chance of being economic for sale to the animal feed compounders. In Table 2.5 there is a list of the operating costs for a number of projected 100,000 t/a methanol SCP plants (only the ICI plant was built). The key constraint is the dependence of operating cost on substrate price, methanol, which is as high as 59 per cent for ICI's Pruteen plant – so the economics of these plants are very sensitive to rises in the price of their feedstock. A writer who reviewed the economics of SCP in 1986 commented that

> Soya meal sells from around $200/ton and fish-meal at $350–400 . . . even at price levels approaching fish-meal, with methanol prices at $150/ton . . . these plants will barely cover their operating costs. Even during the 1960s, at a time when methanol prices were much lower in relation to protein feeds, the economics of such plants were so dependent on substrate that maximum conversion efficiencies were essential and all other costs, i.e. utilities for aeration, de-watering and drying, had to be minimised. . . . When capital costs are considered, the schemes become entirely academic . . . repayment of capital costs and interest even on a $100 million plant would be in excess of $15 million/a.
>
> (Hacking 1986: 100)

These capital costs would add more than $150/ton to the price of the product, and it is this addition that threatens to push the operation into non-viability. This analysis is consistent with the ICI managers' comments on the economics of Pruteen. They knew that their plant was uneconomic after the 1979 methanol price rise, but with the capital costs sunk, or lost, in the plant, their aim quickly became to prove and then license the technology to other countries. One of the

Table 2.5 Operating costs of 100,000 t/a SCP processes based on methanol

Item	% of total cost
Dewatering	19
Off-site services	16
Fermentation	14
Drying	12
Storage and packing	12
On-site services	11
Compression	9
Effluent treatment	4
Raw materials	3

Source: Reproduced from Hacking 1986: 99. Data from ICI as quoted in Fishlock (1982: 104).

Table 2.6 Capital costs of the Pruteen process, 1980

	Norprotein (%)	*CTIP* (%)	*Hoechst/ Uhde* (%)	*ICI* (%)	*SRI Inter- national*[a] (%)
Methanol	51	50	46	59	49
Other chemicals	19	28	20	17	28
Utilities	16	15	23	24	15
Labour	9		4	–	2
Maintenance &		7			
administration	5		7	–	6
Methanol price	550 Nkr t^{-1}	NA	200 DM t^{-1}	NA	107 US $ t^{-1}

Source: Reproduced from Hacking 1986: 101. Data from M. Ericcson, L. Ebbinghaus and M. Lindblom (1981).

Notes: [a] SRI International's evaluation of a process design based principally on ICI patents. NA = Not available.
(The calculations were made before the OPEC price rises of 1979.)

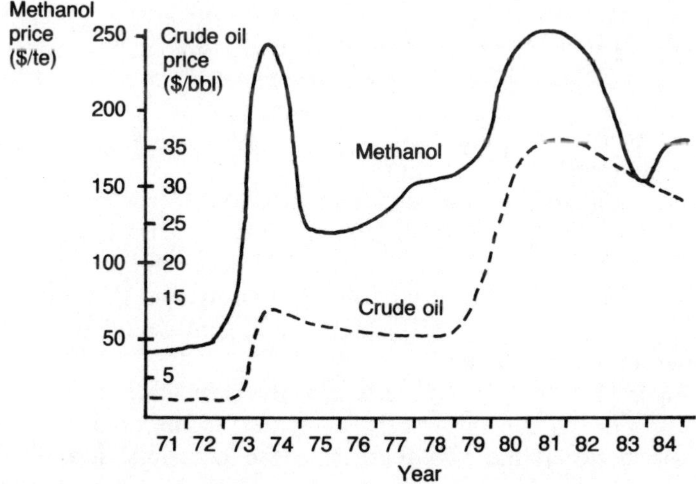

Figure 2.3 Prices of methanol and crude oil (US$), 1971–84
Source: Reproduced from Hacking 1986: 79

managers involved in building the plant thought that in the post-1979 economic conditions, the Pruteen plant could just meet its production costs when operating at near half its rated capacity.

In the breakdown of capital costs for the Pruteen plant shown in Table 2.6 it is clear that no one component of the process exceeds 20 per cent of the total capital cost. In particular, the capital costs of the tried and tested technology for dewatering and drying the product

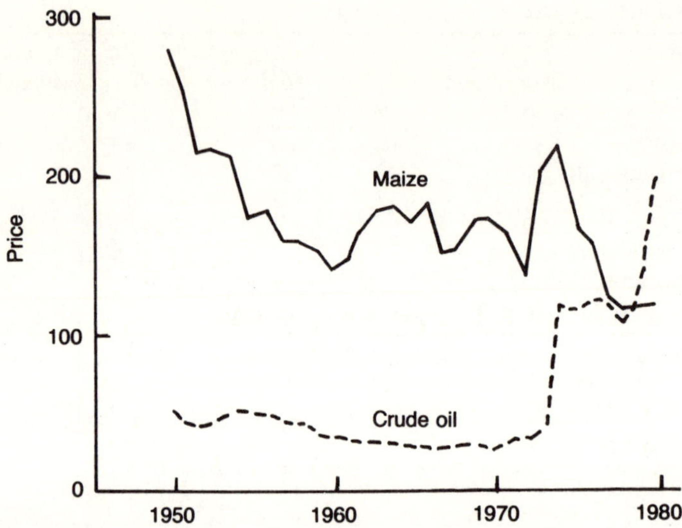

Figure 2.4 Prices of maize and crude oil (US$ 1979), 1950–80
Source: Reproduced from Hacking 1986: 74

are greater than the fermentation and compression equipment costs (31 per cent against 25 per cent).

Figures 2.3 and 2.4 show what happened to the economics of SCP in the 1970s. The 1973 oil shock led to rises in crude oil and methanol prices of the order of six times the original prices. Methanol fell back sharply in 1974 and 1975, crude oil less sharply, but then came the second oil price shock of 1979. Prices eased gently until 1986, when a moderate drop occurred.

In Figure 2.4 the price of maize effectively indicates world prices for bulk protein, and this can be compared to the crude-oil price. The gap between the two represents the economic incentive to convert from one commodity to the other – SCP technology produces protein from hydrocarbons, whereas biomass production and alcohol fermentation (as in Brazil) is the technology that converts food to fuels. The gap between the two, which supported the economic case for SCP, collapsed with the second oil price shock.

However, the substantial fall in the price of hydrocarbon substrates since 1986 has made the economic case marginal or favourable for SCP once more – depending on process design. These changed economic conditions mean that Pruteen would almost certainly be economic once more (as some ICI managers agreed; see Chapter 7), provided it was carefully marketed and if capital costs were written off. The decision not to do so is a strategic one while the plant continues to exist and with the option of recommencing production. The boxes which follow contain summaries of the key elements in each company's development of the new technology.

Box 2.2 Main events in the BP story

1957	Two-year contract for microbial cleaning system for BP France Lavera Refinery awarded.
1959	Champagnat, head of BP research, is now interested in other microbial technology applications. He develops a programme of research in conjunction with French government microbiological research centre. Both gas-oil and n-alkane routes found to be feasible on a laboratory scale.
1960	BP decides to build pilot plants for both processes, one in the UK, one in France. The French are given the choice of process and choose gas-oil.
1963	Lavera pilot plant begins operation. Champagnat and his research team publish the results of four years of work in *Nature* and a general upsurge in interest in SCP begins.
1965	Grangemouth pilot plant begins operation.
1965–8	The original idea of using the processes to remove n-alkanes from waxy Libyan crude and produce human-grade yeast gives way to the idea of an animal-feed market.
1971	Grangemouth 'semi-industrial' plant begins operation with capacity of 4,000 t/a, based on n-alkanes.
1972	Lavera semi-industrial plant in operation, nominal capacity of 20,000 t/a. Venezuelans offer to buy full-scale n-alkane plant if BP can run Sarroch, the full-scale plant, without problems for six months.
1973	First oil price rise.
1975	BP now assume oil price rise permanent.
1976	Lavera 20,000 t/a plant closed on economic grounds after two or three years of production and sale of product on the French animal-feed market. The 100,000 t/a Sarroch plant is completed.
1978	Sarroch plant dismantled, BP compiles its ten-volume work containing details of all their technical work. BP have yet to sell this, at £100,000.

Box 2.3 Main events in the ICI story

1968	A business study on diversification prospects for Agricultural Division suggests protein from methane, the suggestion is to produce pig feed.
1969	Ag Division switch to methanol as a substrate.

<table>
<tr><td>1972</td><td>A 1,000 t/a pilot plant in operation. ICI becomes aware of the skimmed milk replacer market for veal calves. Pruteen business area formed.</td></tr>
<tr><td>1973</td><td>ICI main board rejects first application from Ag Division for commercial-scale plant of 100,000 t/a capacity.</td></tr>
<tr><td>1974–6</td><td>Board members visiting Ag Division to examine proposals for commercial-size plant.</td></tr>
<tr><td>1976</td><td>Main board approve 70,000 t/a plant, to be built by John Brown Engineering.</td></tr>
<tr><td>1979</td><td>Commercial Pruteen plant commissioned.</td></tr>
<tr><td>1980</td><td>Pruteen plant runs at up to 80 per cent capacity for short campaigns. Product is sold as veal calf milk replacer. Price of natural gas has risen rapidly in 1974–9 period, undermining pig and cattle feed market prospects.</td></tr>
<tr><td>1981</td><td>Research shifts into human food uses.</td></tr>
<tr><td>1983</td><td>Negotiations with Japanese food companies for sale of nucleic acid fraction of Pruteen. Another pilot plant is planned for fractionation of Pruteen.</td></tr>
<tr><td>1984</td><td>Japanese withdraw from negotiations. Pruteen business area merged with some Ag Division research to form Biological Products business. Pruteen plant still stands, unused.</td></tr>
</table>

Box 2.4 Main events in the RHM story

<table>
<tr><td>1964</td><td>Research begins at instigation of Lord Rank, on starch-fungus fermentation technology.</td></tr>
<tr><td>1965</td><td>Animal trials begin on C1 organism.</td></tr>
<tr><td>1966</td><td>Product development begins as a Third World protein supplement.</td></tr>
<tr><td>1968</td><td>RHM switch from batch to continuous fermentation.</td></tr>
<tr><td>1969</td><td>Market orientation changes from Third World to First World product.</td></tr>
<tr><td>1970</td><td>Change to new A3/5 organism, a fusarium. Patenting activity starts. Pilot plant begins production.</td></tr>
<tr><td>1974</td><td>Human nutrition trials begin. Scale-up studies begin.</td></tr>
<tr><td>1976</td><td>Food science input to research team begins.</td></tr>
<tr><td>1978</td><td>Submitted to MAFF for approval as a human food.</td></tr>
<tr><td>1980</td><td>Preliminary clearance from UK government.</td></tr>
<tr><td>1984</td><td>ICI and RHM decide to work together. ICI's pilot Pruteen plant is used for mycoprotein fermentation. Sainsbury's agree to work with RHM on product formulation. 'Savoury Pie' product goes on sale at key Sainsbury branches.</td></tr>
<tr><td>1985–9</td><td>Marlow Foods, jointly owned by ICI and RHM, is responsible for the production of mycoprotein. 'Quorn' is sold on to food manufacturers for incorporation in an increasingly wide range of precooked dishes.</td></tr>
<tr><td>1989</td><td>Number of dishes on market now approximately 40, on sale in numerous supermarket chains. Marlow Foods now considers what investment to make in increased plant capacity, e.g. whether greater than 5,000 t/a.</td></tr>
</table>

Box 2.5 Main events in the Liquichimica story

1971 Pilot plant built.
1973 Construction begins on three-plant complex.
1975 SCP complex completed, Reggio di Calabria.
1977 President of Liquichimica files for bankruptcy, Italian government appoints liquidator.
1978 ENI buys Liquichimica plant, operates original Liquichimica n-alkane refining business as before, three-plant complex closed.

3 The markets and the environment during SCP development

This chapter examines the dynamic role of market ideas in the development of projects, where any idea of how the product or process technology might be used by groups outside the firm is counted as a market idea. Also, this chapter analyses how the companies conceived the environmental influences on the SCP processes – that is, public opinion, economic forecasts and the behaviour of governments. At the back of all market conceptions for SCP products was the belief that human population growth in the lesser-developed countries would outstrip those countries' ability to feed themselves in the near future, that is, during the 1970s. Allied to this idea was the idea that protein would be in particularly short supply and that therefore there would be a 'protein gap'. There was thought to be a need for new forms of protein that could be supplied to the human food chain. All sorts of potential new protein sources were investigated, of which SCP was only one example; others were the extraction of protein from leaves and the vigorous development of new strains of crops (see Stanley 1977).

The original market conception that led to the start of research often differed greatly from the sophisticated market understanding that eventually developed. However, three basic market concepts emerged and even if the target market changed on a project, it changed to another one of these three basic market ideas:

1 Market the protein as an animal feed in the First World.
2 Market the protein for direct human consumption, First or Third World.
3 The 'political' market, where Third World governments would license SCP technology to reduce their soya imports or to improve the nutrition of their own people.

The potentially most profitable way of making money out of this situation was the production of SCP for the animal feed market, which is typically made up, or 'compounded', from soya-meal and fish-meal, both produced in quantity in the Third World. If Third World governments ever begin to retain their soya crops to feed their own rapidly growing populations, this would raise animal feed prices in the West and create a growing and long-term market for SCP (Hamer 1979). This was the market basis of the hydrocarbon-based SCP projects, which before the oil price rises of the 1970s were economic propositions with a rosy future.

THE WESTERN ANIMAL FEED MARKET

The feed compounders

In contrast to the tropical countries of the Third World, Europe has an organised feed industry where the feed compounders – companies like Dalgety and Spillers – make up feeds for different types of animal and distribute it to the farmers. They buy in soya-meal and fish-meal from the producer nations, the United States, Brazil and Argentina, then vary the proportions of the two meals in the feeds they sell. They minimise the cost of the final feed while keeping the feed nutritionally suitable for the species and age of animal for which it is intended.

Soya-meal is the base of most of their animal feeds, but soya alone is not nutritionally balanced for animals. It is low in the proportion of sulphur-containing essential amino acids (i.e. lysine and methionine). The animals would have to eat an excess of soya-meal to obtain the amino acids they needed for rapid growth, so it becomes economic to add smaller amounts of fish-meal, which is more expensive than soya, but is rich in lysine and methionine (see Chapter 2).

The animal feed market varies from small-volume, high-value markets, to the bulk markets with low value added, so there is a roughly inverse relationship of price to volume. The smallest markets carry the highest price per tonne, the largest have the cheapest prices per tonne. The higher end of this range starts with feeds for goldfish, then the laboratory animal market, then the calf milk replacer market. The lower end contains the bulk cattle, piglet and broiler feeds. At the top end, the feeds sell for two to three times the price of soya-meal, ranging close to soya-meal price for the lower end of the feed market.

So SCP products would not be total replacements for soya-meal. To the feed compounders they would be an alternative source of protein which, depending on their lysine and methionine content, would have a value somewhere between that of soya-meal and that of fish-meal. The importance of this complexity of the feed markets is that the simple equation for the viability of SCP has partly broken down, that is to say that the simple comparison of the price of a tonne of SCP with the price of a tonne of soya-meal is no longer valid – the feed can be marketed into one of the constituent animal feed markets at a price above that of soya, depending on its quality characteristics.

The value of SCP to the compounders

Besides lysine and methionine content there were other characteristics of SCP that gave it a greater value to the compounders than soya-meal. When chickens and piglets eat too much fish-meal their flesh carries the taint of the fish, so that it can be tasted. As far as the public value their bacon rashers free of this taint, they will pay more for the bacon and the farmers will pay more for a feed free from fish-meal. SCP products gain in value as a fish-meal replacement since they do not contain the fishy taste compounds. Marks & Spencer are a company that involves itself in the specification of the feeds for their meat supplier's animals and they were prepared to pay a premium for this 'no-fishy-taste' quality of SCP feeds.

> One of their producers bought 8 000 tonnes from us in one year. If the economics had been right they would have liked to have moved all their suppliers of turkey and pig meat to using Pruteen instead of fish-meal . . . it was a very highly regarded product.
>
> (ICI manager)

This manager estimated that the less cost-conscious companies like Marks & Spencer would have been prepared to pay maybe 2.5 times the price of soya to put 4 per cent Pruteen into their feeds.

One of the tests in the ICI nutritional programme showed that chickens and fish fed on a part-Pruteen diet grew faster than on conventional feed mixes, which was again used by ICI to argue that Pruteen had a greater value to the compounders than competitor feeds.

ICI's understanding of the feed market was largely obtained from feed compounders, and this understanding gave them greater precision in the pricing and economics of their product. The identification of tiers of value in the feed market and especially of the important veal calf milk replacer market, gave them the choice of building a plant whose capacity matched the demand of the higher value tiers.

The veal calf milk replacer market

After the double pinch of the oil shocks and rising soya-meal prices, ICI found that the economic equation left them unable even to cover marginal costs when they sold Pruteen into the chicken and cattle feed markets. Pruteen was driven out of these low-value, substantially soya-based feed markets to the higher-value ones – the principal higher-value market was the veal calf milk replacer market. Not only ICI, but BP, Bel Industrie and Dansk Bioprotein have seen this market as a premium market of sufficient size to support SCP production at a moderate scale in times of high oil prices. Dansk Bioprotein based their original business plan on this market alone.

Veal calves have a lifetime of about 20 weeks and during this time they would normally feed on their mother's milk, but this milk is taken and used for human consumption. The farmers then use a high-quality milk replacer feed for the young animals, which is of higher value than the chicken and cattle feeds. It also has to be free of fish-meal since the delicate taste of veal is easily spoiled, and of iron, which would allow the meat to colour brown rather than white. SCP has here the perfect market, fussy about just those characteristics that define SCP as a rival product to fish-meal.

In the late 1960s the milk replacer industry was using protein isolates obtained from soya (soya 'milk'), but this changed in the early 1970s as the EC began to accumulate huge quantities of dried skimmed milk as a result of guaranteeing high prices for fresh milk to dairy farmers. In the early 1970s the Community began subsidising alternative uses for skimmed milk to reduce its stockpiles, rather than reduce the quotas for fresh milk production. One of the uses was as a calf milk replacer, because it was, naturally enough, nutritionally ideal for raising calves. So, in order to support small-scale dairy farmers, calves were deprived of their mother's milk, the milk was expensively processed so that it could be stored for years in ware-houses as dried skimmed milk, then it was subsidised so veal calf farmers would make it up into milk and feed it to their calves. The Community even made it mandatory for veal calf farmers to include a proportion of skimmed milk in their feeds, to make sure the skimmed milk mountains were run down.

Pruteen and Toprina were nutritionally similar to skimmed milk powder and in theory were far cheaper, but if the price of skimmed milk powder was kept low to encourage its consumption, the SCP products could not ask for a greater price. Where skimmed milk powder would normally sell at over £1,000/tonne, a price commensurate with its use in human foods, the EC kept the price below £500/

tonne from 1974 until 1987, the critical period for the SCP producers, who all had plant coming on stream in this period.

The BP Lavera plant successfully sold much of its output into the French skimmed milk replacer market for several years, showing that farmers and compounders would buy SCP in France, but after the oil price rise this semi-industrial-sized plant was making a product which cost twice as much as subsidised skimmed milk powder. Bel Industrie estimated the size of the French calf milk replacer market at 50–60,000 t/a, of which 10 per cent was a potential market for SCP products.

In 1986–7 the EC at last began to cut its dairy quotas and so started to tackle the problem of overproduction of milk and removing the need for a subsidy for skimmed milk. The result was a doubling in the price of skimmed milk within a year, to around £1,300/tonne, a price which knocks it out of the market as a veal calf milk replacer. A Bel Industrie manager noticed that the price of veal meat was rising because of this – by June 1988 it had risen by 15 per cent from the year before and the veal calf producers were looking for cheaper feeds. Bel and other companies expected the EC subsidy to be cut again, but here they were trying to make sense of the political movements within the EC. Outside groups were campaigning to have the animal feed subsidy cut further – in 1988 the Biscuit, Cake, Chocolate and Confectionery Alliance were protesting to Brussels about the EC continuing its policy of 60 per cent subsidies on skimmed milk sold for animal feed at a time when skimmed milk was running short for human food uses, because of the severity of the cuts in milk quotas. The significance of these recent changes is that there is once more (1993) a niche market for SCP products in the form of milk replacer products.

THE EXTENT TO WHICH MARKET CONCEPT INFLUENCED EVOLUTION OF PROJECTS

The technology and organisation of the SCP projects were influenced by the developing idea of a market. The projects would not have begun unless some market concept was in place, but as shown in the last section, the market idea developed complexity early in the projects, and later it will be seen that it sometimes changed fundamentally in the lifetime of the projects. The next sections examine how market ideas influenced choices in research and development.

Idea of market influences the choice of substrate and organism

RHM intended their protein product to be a food for human consumption, and managers explained that RHM, being a food company, understood what people are prepared to eat and that this understanding influenced their decision to use only food-grade materials at all stages of the fermentation.

The company's attitude in relation to the choice of a food-grade substrate was described:

> we have a prejudice, perhaps, that people don't eat food that you make first of all for animals and they don't eat food that is manufactured from petrochemical residues, or surplus sludges or this that and the other. (RHM manager)

The choice of a fungus rather than a bacterium or yeast was also based on assumptions about how people would perceive a bacterial or yeast product. The idea of eating bacteria was seen as unacceptable to most people, although both yeasts and fungi are eaten by people, RHM chose a fungus because it had the possibility of being marketed as a type of mushroom. Filamentous fungi, unlike single-cell organisms, could also be given texture, which in RHM's opinion could be made similar to the texture of meat. An RHM manager deliberately distinguished the RHM product mycoprotein from SCP, because SCP 'is either powders or creams or some degree of wet growth that renders it only suitable for animal consumption'. He also made the point that it was biologically incorrect to refer to mycoprotein as a single-cell protein, since it was formed from filaments of cells connected one to another. In all this reasoning over the choice of a substrate and an organism RHM did no market research. Their choice of substrate and organism was based on their knowledge of the human food market, which was obtained from their experience with other food products.

A manager in Hoechst gave a detailed example of the factors influencing Hoechst's choice of organism which were similar to those influencing ICI, BP and Liquichimica. After the initial screening of the organisms they had collected, Hoechst prepared a short list of four organisms, which were then compared not only for various technical characteristics such as speed of growth, but also for nucleic acid content and the proportions of the different essential amino acids. These last two characteristics were only important if there was some understanding of the market for animal feeds, and the final choice of organism was one which did best against these two 'market-technical' characteristics as well as other more technical characteristics, such as speed of growth. So at the very start of the research process market influences were brought to bear on process design.

However, one of the ICI managers felt that ICI was lucky to have an organism with such a very good balance of protein, energy content, minerals and amino acids. By the time ICI had discovered precisely the ideal organism qualities, they had already chosen the organism on a cruder matrix of market assumptions and begun to build plant adapted to this. This was because the development of their market understanding proceeded at the same time as the technical development of the plant and some basic choices (such as which organism to choose) were forced early on in this development.

MARKET IDEA INFLUENCES THE CONSTRUCTION OF PROCESS PLANT AND PRODUCT

Expected size of market affects the planned size of plant

The size of the plant depended on some market-size assumptions. The first and obvious one was that the total animal feed market was huge, with a consumption of millions of tonnes in Western Europe alone, so the plant could be as large as was technically efficient if its output was competitive in all of this market. But if SCP was only going to be competitive in the skimmed milk replacer market, the size of the plant should be governed by the size of this market. There was a debate within ICI on this issue when the plant was approved by the main board in 1976. A number of managers believed that had a smaller plant been built with a high proportion of output going to the skimmed milk replacer market, it would still be running profitably today. The capacity of the smaller plant would have been perhaps 25,000 t/a with a constant 15,000 to 20,000 t/a going to the skimmed milk replacer market. But the larger plant was built instead, for a combination of reasons (see Chapter 6).

Animal feeding characteristics affect the design of the product

The different feed markets required slightly different products. The 'granular' form of Pruteen was for the chicken feed market, while the calf milk replacer was a little more complex to prepare for the market. It needed to be low in iron and the fermenter needed to be run within tighter limits. In fact, what amounted to a mini programme of research was mounted to modify Pruteen for this market. An ideal particle size was picked and a wetting agent found to prepare the product for mixing into skimmed milk.[1] ICI found that the kind of dust they were producing by grinding Pruteen could cause adverse effects in some people, so they began adding cooking oil to prevent the dust becoming airborne. The construction of the post-fermentation processing equipment also depended on choices about the proportions and volume of Pruteen that would be consumed by each market type.

Two inadequacies of the product for its market remedied through the use of nutrition science

Another example of how the process and product were tailored to fit the market requirements was the use of selenium. This is a trace element essential, in different quantities, to the growth of most organisms. At first ICI added to the fermenter the amount necessary for optimal growth of their bacterium, but a problem arose because this was insufficient selenium for the animals that ate Pruteen as a replacement for fish-meal in nutrition tests. Fish-meal had happened to contain the right amount of selenium for animals so the compounders had not had to quantify the feed requirements for this element; their industry had learnt how to compound ideal feeds through trial and error over decades and the industry was only cognisant of the differences between standard feed inputs. ICI's strategy of talking to the compounders had not therefore revealed selenium as a critical component of feeds. Animals fed on Pruteen instead of fish-meal suffered selenium deficiency and failed to grow rapidly. ICI's research department isolated the problem, quantified the selenium need and added extra selenium to Pruteen to make up for its deficiency. Here, it was the use of nutritional science and analysis of Pruteen and fish-meal that gave an increased precision to ICI's understanding of the market – the understanding that was 'science' was informing the market understanding.

The market requirements also affected the sourcing of the phosphate rock, which was one of the nutritional inputs to the fermentation. ICI learnt to be careful about the source of this rock because some contained high proportions of heavy metals, like cobalt, which were poisonous to animals. It was a matter of determining the levels of heavy metal that could be tolerated by the animals and ensuring no rock was used that contained greater amounts than this maximum. As with selenium, this was a problem that did not occur when fish-meal was used as animal feed, because fish-meal contains sufficient phosphate and its phosphate is not associated with heavy metals.

So nutritional evidence on animals' needs became incorporated into the product formulation. This can be seen as a process by which the market needs were in process of being refined and articulated by nutritional science and expressed in nutritional science terms. It was not a matter of collating market needs as expressed by users and delivering them to the process engineers or product formulators – that market concept was a product of the experience of feed compounders with traditional feeds. Innovative feed sources such as SCP required a more precise articulation of market need.

MARKET CONCEPT AFFECTS RESEARCH

As much as a third of all ICI research expenditure was on toxicological and nutrition research. ICI were aware that public opinion had probably been the principal cause of the failure of SCP in Japan (see Appendix) and Italy, and this heightened their sensitivity to possible accusations of product toxicity, especially since theirs was a bacterial product. The result was an unusually thorough toxicology and nutrition research programme. The toxicological research manager felt that this perceived need to go further than satisfying purely scientific criteria had a major influence on the way in which the toxicological research was organised. He remarked that

> instead of saying, we'd like to do a few things with chicks and pigs and things, we decided what the markets with best value were going to be, we talked to the companies about the trials and evidence they needed. We had the trials done across the world and received evidence from companies doing their own trials. We put together a package of its being a very efficacious product.

This manager adopted the policy of working closely with the Department of Health from the beginning of the project, with the explicit aim of building the Health department's trust in ICI's methods and claims for Pruteen. ICI did this by being unusually open about the results of their toxicology trials. They revealed results with possible negative interpretations – the company revealed its internal results that showed animals could die when fed a part-Pruteen diet, as well as thrive on such diets. They could show that ICI understood the different responses to the diets and that they could correct any deficiencies by adding the necessary nutrients.

ICI considered using Pruteen for human food in 1973–4, before plant construction was approved. However, human food research was seen as a major diversion of research effort, so that once a choice had been made to produce for the larger animal-feed market a decision was taken to stop research from diversifying into human food possibilities. In this way the market conception was supposed to restrain the research possibilities, while the product was also constrained by its process of manufacture to be an animal feed, not a human food.

Change in team structure changes market conception

The original proponents of research into SCP within BP dominated the early stages of the project in France. These two product champions believed, like so many others, that SCP would solve world hunger, and until 1966 the company was still thinking in terms of food for human beings. This was a period when BP advertised the 'petrol into beefsteaks' idea and published photographs of Champagnat, one of the leading research scientists, eating Toprina (Laine, Snell and Peet 1976). By 1968 the idea of the feed market had become dominant. The change had been gradual and occurred as more people with a variety of backgrounds joined what had been a small-scale project. Some of them were acquainted with the feed market and they shifted the thinking away from the conception of a human food market. So, one of the people who joined in 1966 was an agronomist and because of this background argued that it would never be possible for the yeast to be eaten by people. By 1968 the new idea of the animal feed market was the guide for BP research, and they too were developing a high-protein fish-meal replacer for animal feed.

Technology before market

The SugarCo research programme generated many projects involving fermentation. The citrus waste and carob projects which were taken furthest are discussed in the Appendix. There was also a process developed for the fermentation of high-sugar residue wastes, but once developed no market was found to exist and no one wanted to buy the technology. They also developed a very simple fermenter which was a hole in the ground lined with concrete, fitted with a stirrer and into which they bubbled air. This was abandoned after about six or seven months of work because the project manager could not see how they were to get money out of it.

The approach was to develop a precise technology and then to look for a market niche, which rarely existed. The only success came about through a series of happy coincidences – this was their process for using waste carbohydrate solutions. The fermentation projects director thought this might match the needs of a new sugar-refining process in the SugarCo Technical Services department. At the same time as he approached the sugar-refining division, Bassetts, the Liquorice Allsort manufacturer, was also talking to SugarCo Technical Services about a sugar effluent problem. So Technical Services used Bassetts as a test bed to develop the research department's technology into a full-scale working plant. Bassett's paid for it, but SugarCo never succeeded in selling it anywhere else. The problem Bassetts had with disposal of confectionery waste in the River Don would appear to be unique in Britain – such sucrose-rich effluents are rare. A paper written about the process shows that the investment was a marginally less costly solution to Bassetts than using a conventional trickle-digestion plant (Hacking 1986).

The research department's idea of the market

Occasionally managers widened their comments about the market to generalisations about how far the research department could use market ideas. A research manager remarked that,

> you never completely get rid of the idea of a commercial end product from the research. However divorced the scientist is from the market, he still thinks that one day this may be something amazing, it's that crude, but there's still an element . . . that's industrial research after all.

The marketing managers tended to be caustic when discussing the role market ideas play in the R & D department:

> When they are telling you the stories of the great inventions, they will tell you, well, of course we thought through the marketing issues, and as a result we decided how we were going to do it. A load of lies, most of them found something and in a process of moving forward, came to understand the issues.

In response to the suggestion that the scientist in R & D would have some kind of image of what they were doing as being a useful product, this manager commented on his own R & D department:

> Not at all. The R & D department would not have understood what a market was if it came up and bit them. It was only after

the first 15 years that they hired anyone who could spell 'marketing'. And he sat and pulled together certain figures that were irrelevant to the real challenge . . . at least it was a recognition that somewhere along the line you have to sell the stuff to someone.

This issue of how far market ideas are present in R & D is one of degree – the research scientists could always come up with some 'concept of use' to justify what they were doing. The issue, of course, is whether some outside entity will pay for the new product – is the 'use' guiding technological development one that people will pay for? Even in SugarCo's concrete hole-in-the-ground fermenter, there was a 'conceivable use' which acted as an initial market idea. It is difficult to imagine the development of any technology without a 'use concept', the problem for the firm is in deciding whether the 'market use concept' is a 'market need concept' – will the anticipated use be one that outside bodies will pay for? It is the blurred nature of this difference that allows research departments to develop products for which, in retrospect, it will be decided that there is no market. It is too easy to dismiss research efforts, in retrospect, as 'not being market led' or the equivalent 'put down' which is 'technology-push'. Marketing managers are far more interested in real, target customers and the process of relating existing market characteristics to product development, the process by which technical choices are guided by market characteristics. There is a dialogue between the two during product development and it is its quality that would appear to be important.

FORECASTING THE PROTEIN MARKET

The economic viability of the hydrocarbon projects depended on the relative prices of the hydrocarbon substrate and soya-meal. The high dependence of operating costs on the price of hydrocarbon substrates meant that company forecasts for their soya and hydrocarbon price evolution were critical to the decisions to construct plant.

Expectations for the evolution of hydrocarbon prices

There were fewer management comments about the oil price rises of 1973 and 1979 than about the evolution of the market for soya-meal as an animal feed. They accepted that the 1973 oil price hike had been totally unpredictable. The rise had been counter to all the predictions of economists and the price trends of the last 20 years. One manager pointed out that in real terms, oil had been stable and even become cheaper throughout the 1960s. The 1973 price rise was 'totally unpredictable, totally out of the blue. They blew the world economy, and SCP was one of the casualties. No question of that'. Another agreed that economists were predicting price rises for crude oil in the late 1960s, early 1970s, but not on the scale or with the suddenness of the actual jump, which was to $35 a barrel in 1973. One explained that many predictions were based on the experience of a six- or seven-year cycle in oil prices and the expectation that this would continue in the 1970s. The failure to predict the rise amounted to a failure to predict the political success of OPEC in forcing the rise.

No manager at any time suggested that the economics of SCP estimated pre-1973 could have been improved with regard to the crude-oil price predictions. The expectations in the case of methanol were a little more complex.

At the time that ICI were deciding to build a large-scale plant, in 1976, the price of methanol was still relatively low – it had settled down to double the price pre-1973. This was because it was manufactured from North Sea gas, which was available in large quantities without an equivalently large market – much gas was being flared off as oil-fields came into production. So North Sea gas was cheap.

ICI had developed its unique low-pressure catalytic process for converting natural gas into methanol in the 1960s. The late 1960s to early 1970s had seen a steep fall in the price of methanol as this method of production had become standard. In the early 1970s North Sea gas was not yet used as a fuel and so its value was not strongly linked to the price of crude, which represented the price of energy. The result of this decoupling was that the 1973 oil price rise did not severely affect the economics of methanol-based SCP processes such as ICI's – indeed, methanol-based processes were seen to be the way forward in SCP conferences during this period. Once the UK switched over to using natural gas instead of town gas as a fuel, this was no longer the case – the major use for natural gas became that of a fuel. In the later 1970s the price began to rise in real terms (see Fig. 2.3), and in the 1979 oil price jump the price of methanol doubled because of its dependence on the price of natural gas and the increasingly strong link between the price of natural gas and the price of energy (crude oil). It was this second jump in price that made Pruteen uneconomic to produce.

ICI had thought the margin of profit in the early 1970s to be so large that any 'reasonable' methanol price rise, of the order of tens of per cent, would not affect them. They did not predict so large a rise and they could only have done so if they had been able to predict the opening-up of entirely new markets for natural gas that made natural gas an internationally traded product with a standard price. This occurred as countries like the UK converted to using natural gas as a domestic fuel.

In this situation the use and interpretation of quantitative forecasting techniques were not as important as an ability to predict the radical shifts in structure of the hydrocarbon and soya-meal markets.

Evolution of the price of soya-meal

SCP was to compete as an animal feed and the price of animal feed was linked to soya-meal prices. Even the niche markets which allowed a high value added per tonne of SCP product were linked to the base soya price. If soya was cheap enough, compounders could forego some of the higher-value feed inputs and simply use more soya in their feeds.

The next sections look at managers' reconstructions of how they saw the protein or soya price evolving during the 1960s and 1970s, the different interpretations of the 'blip' in prices in the period 1973 to 1975 and the hindsight reasoning which they use to explain the failure to predict the price changes.

The protein gap

One of the Shell managers thought that a shortage of animal feed protein began to make itself felt by the early 1960s. He talked of a general worry about the adequacy of protein supplies in general and the level of resources that the earth could provide. This was partly fuelled by the Club of Rome report which received a great deal of attention at a time when there was a fashion for world futures studies, in the 1960s and 1970s (see Cole 1978 for comments on this report and others) most of which forecast dire food, energy and materials shortages within 10 to 20 years.

Marketing and toxicological research managers in ICI explained that ICI thought that there would be more people who would want to eat meat as living standards rose in the third world. At the same time,

> there were finite supplies of existing feeds . . . scares in the early 1970s over fish-meal supplies, shoals disappearing for odd years . . . soya seen as being limited by a finite amount of land surface,

fairly specialised in its needs, unlikely to be able to respond to the demand for animal feed.

At this time the price of fish rose due to low catches off the Peruvian coast, while the US soya harvest was also poor for one or two years. These events, concurrent with the oil price increase of 1973, led to a trebling in the price of soya-meal.

The rise in the price of crude oil undermined the economics of SCP; the rise in the price of soya improved them. For ICI the rise in the soya price was crucial because it came when they were deciding whether or not to invest in a full-scale SCP plant. One manager expressed what he felt was the Agricultural Division view:

> There was no reason whatsoever to suppose that protein prices were going to fall. The margin that you could get on this process at the time was very high. And as far as you could look forward it was going to stay high.

ICI based their estimates of the Pruteen project's economics on this high price for soya, while at the same time the price of methanol had not risen as sharply as crude oil, so the project looked especially attractive at a time when the alkane-based projects were in doubt. But the soya price fell back to lower than its pre-1973 level over the three or four years after 1974. It has remained at this low level until the present, and this drop, with the second doubling in methanol prices in 1979–80, ruined the economics of the Pruteen process completely. An ICI manager made the bleak comment that 'Protein prices have continued now for five or six years at what is something like a 50-year low in real terms. We got that wrong'. The Agricultural Division view of the initial rise in soya prices was that it could be seen as the start of the long forecast protein shortage. Other managers, in RHM and SugarCo, all expressed the view that the soya price rise could have been seen to have been a temporary 'blip' after which it would return to its normal level. One of these commented that RHM had been buying soya for 50 years and everyone in that industry expected the price to fall back again. Discontinuities in price were possible, but

> if you are going to postulate a plateau change in something which has been level for 40 years, you have really got to advance a better reason than, it just happens to be up there just now. That seemed to us to be what they [ICI] had done . . . our people were saying there is nothing in the agricultural system that is driving it to stay up there.

Perhaps most interesting is the view of others in ICI, involved in monitoring world trends in cereal prices. A manager in the Plant Protection Division at Fernhurst described the evolution of the soya market in this period:

> in the late 1960s and 1970s there was a period when demand was increasing slightly faster than supply. This led to a steady tightening of the supply–demand balance. A gentle rise in prices in real terms. Then when the oil crisis came there was an outbreak of paranoia and the prices of all commodities rose many times higher than could have been predicted. In turn, this overreaction gradually evaporated and the price drifted back again. But this had stimulated increased production in Brazil and the United States. . . . A period of overproduction began, a period for the late 1970s and early 1980s of weak prices.

For these people, the Club of Rome report and all the protein shortage forecasts of the 1960s and early 1970s were based on an extrapolation of this first period of gently increasing demand over supply. It was felt that, although the extreme jump in prices in 1973 was not predictable, once this had occurred the increases in soya production and the consequent drop in price were predictable. The understanding that prices would drop back depended crucially on the view that there were vast areas in Brazil and Argentina that could be turned over to soya production if there was a price incentive to do so. Supply was able to respond to increased demand, and this would prevent any permanent price increase. This view was resisted by those controlling the Pruteen project.

The Agricultural Division managers had a variety of explanations for why they failed to predict the return of soya prices to their postwar norm. One suggested that it was partly ICI's invention of new crop sprays and the introduction of the no-till farming technique to Brazil that helped increase soya production in the late 1970s, early 1980s. Since such qualitative changes were difficult to forecast, the Agricultural Division's view that soya prices would stay high or continue to rise would appear more reasonable, but in a conversation with ICI Fernhurst this was discounted as no more than a marginal explanation for increased supply and consequent price falls.

In retrospect, a Pruteen manager admitted that one of the project group's core beliefs that prompted their optimistic views on soya price levels was at fault: 'Basically I think what we didn't believe was that the world would allow so many millions to go hungry and to starve. We got that wrong'. Managers from BP, ICI, Shell and John

Brown pointed to the subsidised production and export of soya by the United States as a reason for its low price:

> the farmers in the USA were lobbying to ensure that production of soya was kept up high with subsidies to farmers. And that suited the American government very well because they, for political reasons, are happy for a number of countries to be dependent on them for protein supplies.

However, since the level of subsidy did not change radically in the critical period for ICI of 1973–6, this has no explanatory value for the progress of the soya price in this period. Of course, if the United States ceased to subsidise soya production, prices might rise, and that would help the economics of SCP. The interpretation of the soya price rise of 1974 as a 'blip', not the beginning of an upward trend, was the right interpretation on a ten-year time scale.

LOSS OF ANIMAL FEED MARKET STIMULATES SEARCH FOR NEW MARKETS

ICI and BP

By the late 1970s BP knew that the plant they had built would barely be economic if its product was sold entirely into the animal feed market, and ICI was in the same position in the early 1980s once their plant was completed in 1979. ICI's initial reaction was to hunt for alternative markets for Pruteen – a strategy BP had already followed during a phase of senior management sponsored diversification in the early 1970s.

By 1980 ICI were looking for alternative markets to the poultry feed market, which was to have taken most of their production. The calf veal market remained viable as did the piglet feed market, but this still left more than half the plant capacity uneconomic. Between 1981 and 1985 the plant was operated in 'campaigns' of full-capacity operation of around 7 tonnes per hour. The produce was stored then ground for the milk replacer market and production was kept low to avoid flooding this market and lowering the price obtained per tonne of Pruteen. A number of ex-Agricultural Division managers thought that ICI could just cover the costs of production with this strategy, despite this being an 'appalling way to run such a large process plant'. The hunt for alternative markets took the form of experiments in fractionating Pruteen[2] and considering it as a collection of chemicals. There were two main fractions, the nucleic acids and a proteinaceous mass. This period was exciting because for a time it looked as if there was a high-value-added market in the Japanese and Korean nucleic acid market. The economics of the Japanese project were thought to be very advantageous partly because of the yen's appreciation against sterling. Negotiations with the Japanese were encouraging to ICI because it appeared that the Japanese had a real problem. They used nucleic acids as food flavourings, and their old source of nucleotide flavourings had been SCP grown on pulp mill wastes. Many of these mills were being closed or superseded by new technology, so the Japanese food flavouring producers were worried about their supplies. The process of marketing ICI's nucleotide proteins to the Japanese was a fairly long process:

> We produced nucleic acid, classic sort of marketing thing, taking samples to the Japanese, four or five times, getting a feel for what they thought of it and then they gradually refined it, could we do this to it, then that and that . . . we met virtually all their requirements. At the last minute they took cold feet.

It was not clear why the Japanese pulled out of negotiations that had continued for a year, but one reason that the marketing manager suggested was a Japanese fear of a repeat of their 1972 experience, when public protest had led to the Japanese SCP projects being banned by their government (see Appendix). If this project had continued, the Japanese would have been openly adding an SCP derivative to human food – although this was what they were doing anyway with SCP grown on sulphite wastes. If this explanation is correct, it would appear that it was just the fact of the Pruteen SCP plant being purpose-built that ruined this market outlet, and the shades of the political row over SCP in Japan reached forward to damage the ICI project.

The failure to secure this market for the nucleic acid fraction of Pruteen ended the chances of profitability for the fractionating project. However, some years of work had been put into developing the market for the proteinaceous fraction which would be left after nucleic acid extraction. This material could in theory be changed in various ways to give the physico-chemical characteristics required by different foods. For a short period it was the job of the research department to investigate how to modify it to give it texture and other desirable human food qualities.

At this point, 1982–4, ICI were thinking of a small integrated plant producing some animal feed, human food and nucleic acids. But the research revealed that Pruteen was a very limited material from which to make human food additives. The proteinaceous fraction of Pruteen did function like skimmed milk, or casein in foods, but one manager described Pruteen as having 'few functional pluses' – even yeast products were found to have more food uses than this bacterial protein product. Yeast products could also be described as 'natural' because the food use was already established, and 'natural' was a word wanted in product descriptions. There was a problem with the government's Trading Standards Board, who told ICI that Pruteen-derived products would have to be described as 'bacterial protein' in any list of food ingredients. This occurred because when Pruteen was developed the form of words with which the product was routinely registered with the Board did not appear important since ICI did not then have the intention of marketing Pruteen derivatives into the human food market. So the Board took its own advice and called the product by a name which was technically correct but a food marketing disaster – 'bacterial protein'. ICI did not have to do market research to believe that any product containing 'bacterial protein' rather than 'skimmed milk', would be at a huge disadvantage.

The only realistic use for Pruteen was as a soup additive, a meat extender, a sausage filler – all low-value-added uses and none viable without the sale of the nucleic acid fraction. The Pruteen marketing manager initiated the internal paper that led to the end of the fractionation project – for reasons of poor market pull in the human food area, the unreliability of the Japanese as long-term customers and the poor macro-economic climate.

Shock loss of animal feed market leads Bel to develop human food market

The dairy company Bel Industrie developed an animal fodder yeast called Protibel which was economic in the low end of the animal feed market until 1973, when the market was abandoned (see Appendix). In Bel Industrie's case, it was the tripling of the cost of the energy-intensive aeration, drying and separation of yeast from water that changed the economics for the worse.

Bel Industrie reacted to the crisis by beginning a long-term programme of research into higher-added-value markets for the yeast product. They developed a product for the calf milk replacement market, just as ICI did, calling it Sassiyen. Both ICI and Bel developed a way of grinding the product into the yeast equivalent of skimmed milk powder, which could be dissolved in water before feeding to calves. This was long-term thinking – it took four years until they could begin marketing Sassiyen in 1978–9, and through this period they worked on the development of their yeast product for the human food market. They also refused to meet large orders (3,000 t/a) from customers in single, niche markets, so they could build an array of smaller markets – this strategy was intended to prevent such a disastrous collapse of the market ever again.

Bel has successfully built a human food market, based on output from a second yeast production plant in Chartres. One of the best outlets is the market for dairy-free products, a health market which has only developed since the 1980s. A good example is the market for dairy-free chocolate, where powdered yeast replaces skimmed milk as a chocolate ingredient. Yeast tablets for child diet supplements are another very high-value market, although this is limited as the waste yeasts from the breweries compete here. Other speciality markets were the goldfish feed market, especially in Germany, and Bel thought this had great potential despite developing slowly, because Bel had to win acceptance for yeast as a fish feed. Others were yeast as a vitamin preparation for shiny dog coats, a binding agent in charcuterie and as a feed in Japanese aquaculture.

This was a strategy of adapting the product to specific markets, the long-term development of a high-value SCP market to safeguard continuous production of the yeast biomass. Bel's success shows that SCP can be sold into the human food market provided time and management effort are committed over the long term, although the economics for a lactic waste-based process are different from those for hydrocarbon-based SCP.

PERCEPTION OF A MARKET TREND LEADS TO PRODUCT DEVELOPMENT

The origins of mycoprotein

The RHM project began as just another 1960s protein project intended to feed the starving millions – another project with a use, but no market. According to RHM, Lord Rank himself initiated the project for philanthropic reasons – he thought there was a chance to help the third world at the same time as disposing of surplus starch from RHM's milling activities. This starch was a by-product of the manufacture of RHM's reduced-starch 'health' bread, 'Nimble'. So a RHM manager could comment:

> While the project was classed as research, consideration of markets did not take place in a significant way. It was only after regulatory

approval was obtained and the project was established scientifically that the real marketing began; then, at least, there was the recognition that somewhere along the line you had to sell the product to someone.

In contrast to the SugarCo examples, which grew out of a grand strategy within R & D, the RHM project began because the chairman intervened and gave the project to R & D. It then continued as a research project without hope of real development, because although it was protected by its champions, no one could see how Rank could make money out of it. Then in the early 1980s, the company saw that the developing trend for healthy eating matched their protein product's nutritional characteristics and the project won the full commercial backing of the company.

RHM had already decided to market mycoprotein products seriously when along came a spate of reports by various food and health committees which began to raise the public's consciousness of diet and health connections. This gave the added rationale for buying mycoprotein products that was previously absent. Until the developing 'healthy eating' trend appeared, it was never clear why consumers should not eat the chicken or beef with which mycoprotein could compete.

A market opportunity had been identified but RHM knew the market needed active development. A member of the commercial team in charge of the product spoke of having to decide what attitude they were going to have towards the product:

> We had said we had to be evangelical about this product. If we don't believe in the product enough, if we keep trying to insinuate this product onto the UK market we are going to fail. We will deserve to fail.

Before the test introduction of the first mycoprotein product in 1983, the Savoury Pie, RHM ran a programme of attitudinal and consumer research and began to talk to other companies, such as Unilever, about joint development. This research led them to distance themselves from soya and SCP. They would not even join the clubs and societies that people who made soya joined – mycoprotein must not be seen as a cheap meat 'substitute', nor was it desirable that the historical link to SCP products should survive. Once their own attitude to the product was decided they began an active promotion campaign,

> first among the decision makers, people like the BMA, the universities, the guys the journalists go to for information . . . like

Tony d'Angelo because he's editor of the *Grocer*. We spent time talking to those people, getting them on board, so we had a cascade theory starting to emerge, we had people saying mycoprotein is worthwhile, yes, its very good for you.

The image of the product was extraordinarily important – a marketing manager pointed out that there was a difference between the peripheral purchases in people's lives and what mycoprotein wanted to be, a 'centre of the plate' purchase:

> not putting chicken or beef in the middle of your plate, that's a whole different cultural change. You are going to stop doing that only for some very profound reasons, something as enormously powerful as 'if you go on eating this stuff, it's going to kill you,' and that's almost the message the NACNE, JACNE[3] people were saying.

Mycoprotein contained no fat, had more protein per unit weight than chicken and contained fibre where chicken and beef do not – its nutritional characteristics now had a market value where previously they had none.

'Novelty' as a food characteristic

The novelty of mycoprotein is that until RHM began feeding it to its staff, no human being had ever eaten this foodstuff. The RHM marketing manager thought the last time anyone had done what he was trying to do with mycoprotein was 'Sir Walter Raleigh with the potato'. In ICI the view was that the last time a novel food had been completely accepted by the public was with margarine – over 60 years ago. There was no doubt within RHM that novelty in food was not 'great news':

> they will buy technology in a watch. But if you turn up with a food and say, 'great news, I've got this totally novel steak, what do you think?' in about four nanoseconds you'll be out on your ear. Nobody wants this kind of thing . . . centre of the plate, they look for the most unprocessed traditional product they can find.

In the 'peripheral' food market, whatever manufacturers did to products like ice cream in terms of shapes, colours and tastes, they kept the ingredients to traditional names. This intense conservatism and avoidance of the novel in foods was supported by many managers by reference to what can only be called the food industry's 'disaster anecdotes'. Two of these were BP's advertising for its SCP product Toprina, in the 1960s (described later in the chapter), and the textured-soya, meat substitute fiasco of the 1970s. In the latter case, many food companies such as Unilever and Nestlé had been preparing to launch a variety of soya-based meat analogues, when Cadbury-Schweppes reached the market first with a range of products that openly declared the fact that they were soya masquerading as meat. ICI's food scientists thought that this product had been inferior and had spoiled the market for other companies. None of the soya products were successful because the public had preferred to pay more for real meat. The conclusion drawn by the food industry had been that conservatism pays and novel food sources should be avoided. So in the marketing of this highly novel foodstuff mycoprotein, RHM would avoid all reference to its being 'new' and link it to known, safe and conservative food concepts.

RHM's understanding of the public perception of healthy foods

RHM had detailed knowledge of the market for food, and this naturally moulded their approach to the marketing of mycoprotein. As far as additives were concerned, the less the better, but at the same time there was scorn for how people avoided certain names of additives like sodium caseinate or sodium chloride, but sought out the common name equivalents; milk powder and salt. This was the level at which the public made decisions and it had to be understood – because the decision to buy would be in terms of their thought structures, not a food scientist's ideal. So there was a particular public conception of fat, fibre, sugar and salt which RHM had to follow:

Fat	'Everyone recognises, for reasons they don't fully understand, that fat isn't healthy. It causes you to fall over with your legs in the air sooner than you should do.'
Protein	'Sort of good for you, like fish, cheese and eggs and things.'
Fibre	'"Well, it seems to, well, make you sort of better." They haven't made the connection to cancer of the bowel or of the lower intestine. But they have heard Kellogg's telling them that, particularly bran fibre, is good for them, and, you know . . . it's a good idea.'
Salt and sugar	'Not fully articulated yet, but certainly people are saying, if its got less salt it must be good for me.'

There was a trend for the public to become more aware of the nutritional breakdown of food and to become more precise in its attitude to each ingredient, and this trend was more pronounced in the UK than in the rest of Europe. This trend meant a developing market for mycoprotein if it could be sold as a low-fat, low-salt and high-protein food rather than as a novel food. Together these qualities justified the use of the word 'wholesome', another word to which the public attached great value. RHM spent months working out whether 'wholesome' was the right word to put on the packet of the Savoury Pie.

The pie was formally launched in 1985 and RHM immediately began market research to find out how the pie was perceived by the consumer. The company was able to develop a hierarchy of reasons for buying the pie. In answer to the question 'Why do you buy it?' the first response was 'It tastes nice'. The second response was value for money. On the issue of pricing it was decided that it was important not to price too low, even when it was possible; the idea was to maintain the idea of value for the money spent, without allowing the price to suggest it could be compared to low-value meat products – or worse, something that was not meat at all. An unexpected piece of marketing survey feedback was the advantage of a product that contained no gristle. RHM had not thought this important, if they had thought about it at all, but:

> it's amazing the number of old people and parents who say, when you ask why they buy this product, because when my kids find these funny bits of biological plumbing inside these [other meat] products, that's the end of it.

The Savoury Pie was seen as a substantial success because it was an innovative product selling in an extremely conservative market, a market where firms were reluctant to innovate but very fast to follow other people's successes. RHM were keen to avoid mycoprotein being thought of as a 'meat substitute' – that was how soya had been thought of and marketed and that had failed, but, 'there is still a tremendous amount of market pull, with meat creating the market idea'. It is the analogy with meat that allowed RHM to see an enormous potential market for mycoprotein. The marketing people wanted to get mycoprotein into the burger, steaklet and recipe dish markets, to produce a mycoprotein product wherever there was a meat product.

This is not to say that the product's future is assured. Partly because of the soya meat-substitute fiasco, a senior R & D manager in a rival food company just laughed at the suggestion that myco-protein had a future. Others outside of RHM agreed that there was no consensus on its prospects. This was because, although RHM had so far preserved the image of mycoprotein very carefully there was an 'alter ego' image for the product. It could be seen as a 'fungus', rather than a 'member of the mushroom family', which is how the Savoury Pie packet describes mycoprotein. 'Fungus' has connotations of rottenness, decay and poison. The critical issue is whether some-one will start to 'market' this opposing image of mycoprotein and whether it would stick to the product. Both RHM and ICI feel they are ready with a countering advertising strategy if this should occur. BP in Italy suffered exactly this fate.

Fashioning a company image through selling Toprina

While RHM presented the Savoury Pie as a product that would be perceived as healthy in the terms in which the consumer understood 'healthy', BP tried to use the positive way in which they thought the public would see a novel protein producer to reinforce the public's positive image of BP. BP went out of its way to advertise its involvement in SCP, and in the years until 1966 this meant forging a direct link between oil and human foodstuffs. One manager commented that

> BP were very clearly blowing the trumpet about bridging the protein gap, solving world hunger by making food from petrol. In fact they had pictures of people at petrol pumps decanting soup from the pumps . . . [that] backfired badly in the end. Very badly.

As the project expanded in the 1966–8 period BP moved away from the idea of feeding people to that of feeding animals, but for the public and the media the association with human food had been fixed in their minds. Italian newspapers routinely referred to BP's plant and product as the effort to make the 'bifstecca petrolio', the petroleum beefsteak (Il Sole 1975). BP never convinced Italian public opinion that Toprina was not intended for human consumption and the initial marketing approach became a liability and contributed to the breakdown of trust between BP, the Italian public and govern-ment. The Liquichimica project manager also referred to this BP marketing as something which was hugely damaging to Liquichimica's SCP project because of the way it affected public opinion on SCP.

BP knew there was a public perception of a protein gap; they assumed a new method of producing protein would be appreciated by the public as a contribution towards solving world hunger and they wanted their name linked to that appreciation. They did not foresee that public hostility to foreign multinationals would translate into a whole series of hostile interpretations of their intentions and of their product (see Chapter 5).

THE ATTEMPTS TO LICENSE SCP TECHNOLOGY TO DEVELOPING COUNTRIES

SCP for animal feed might have become uneconomic in Europe, but there was the chance that other countries would have an economic or political case for licensing the technology. One of the managers of John Brown Engineering who worked on the construction of the Pruteen plant thought that ICI moved from the idea of building plants world-wide to licensing as a way of getting back some of the money spent on the Billingham plant. This involved ICI and John Brown Engineering putting together proposals for around 25 countries, proposals which cost ICI hundreds of thousands of pounds each and in some cases as much as £0.5 million. BP were always ready to license their technology, and the companies approached the same group of countries in the bid to make a return on their SCP investments. Kanegafuchi successfully licensed its process to Liquichimica, but neither BP nor ICI succeeded in licensing their processes, despite the intensity and breadth of negotiations.

After the first oil shock BP knew they would not be building a series of SCP plants in Western Europe, and after this date they continued to construct their Sarroch plant because they wanted two plants (the other was at Lavera in France) to demonstrate to prospective licensees that the technology worked: 'We wanted to go on because for some countries economics doesn't mean anything – some socialist countries'. BP, ICI and Liquichimica were thinking primarily of the Soviet Union, which has persisted with a (now uneconomic) alkane SCP programme (see Appendix). There were also political reasons for the Soviet Union and less-developed countries to be interested in SCP, and one of the managers involved in the licensing effort for ICI commented that these countries were interested in

> reducing the import of soya protein so that they become to a greater extent independent of the soya producers. The decision to build such plants became political. In every case where we have been involved in detailed negotiations the political element has been very marked.

Most of these countries were oil rich with a deficit in foodstuffs. They had rapidly growing markets for meat and therefore a growing dependence on grain imports as a source of animal feed. The economics of SCP in these countries would be similar to the economics in Western Europe, since instead of manufacturing SCP they could always export oil and import soya. Such countries include Mexico, Venezuela, Indonesia, Kuwait, Saudi Arabia, Libya and Algeria.

The only alteration to the economics comes from the subtraction of transport costs from hydrocarbon costs, which makes a significant difference when the hydrocarbon is natural gas. Kuwait and Saudi Arabia are too far from the markets for natural gas for its price in those countries to reflect its value as a fuel in industrial markets, given the necessity to liquify the gas before shipping it. Algeria for a time shipped liquified natural gas to the UK, but once pipelines were built to the North Sea gas-fields this became too expensive. The result is that natural gas appeared almost as a waste material in these countries, as it once was in the North Sea. At least, this was the case until it became possible to build methanol plant; methanol production made sense because it is another commodity with a value on the international market, and unlike liquidised natural gas, it would not evaporate on the way to its industrial markets. Shell had done an econometric analysis which showed that after natural gas, plant which used methanol as a substrate stood the best chance of giving a return on investment during the mid-1970s.

A manager in ICI explained the failure of any country to take up SCP licenses in terms of 'disaster anecdotes':

> We were talking to the Mexicans, then they had an earthquake. We then talked to Russia . . . the deputy PM who believed in this and was pushing it forward dropped down dead at one of the Trade Fairs, 1982–3. There was a change of government in Indonesia when we were just about to clinch the deal. We went back to Mexico and then their whole finance collapsed. Virtually everywhere we went, something happened which was really irrelevant but stopped the project.

If ICI was unlucky there were other reasons for this failure which included the cautious and slow approach of the less-developed countries when negotiating with Western firms. A manager in John Brown said many had had bad experiences with licensing Western technology and that 'they have been rooked left right and centre by the West.'

One of the SCP project managers in Shell was critical of the Arab countries' ability to negotiate, and this manager later worked in Kuwait. According to him they had no technological expertise and didn't know what questions to ask at the negotiating table, and they were thereby forced to trust the company, something which they were naturally reluctant to do when that company was trying to sell its technology for a profit. He thought that only Kuwait was able to make a serious independent assessment of the technology, through the Kuwait Institute for Scientific Research (KISR), which critically compared the Shell, ICI and BP processes. This should have made Kuwait a prime candidate for licensing plant, but he thought the political will to buy was missing because Kuwait was a very rich country that did not need to take risks with an untried technology. They also faced the problem that they could only consume around 30,000 t/a SCP themselves, the rest would be exported to Saudi Arabia. This they were not interested in doing since the main aim of the plant would be to substitute for imports of feed, not to subsidise the Saudi Arabians (they were only being offered licences for 100,000 t/a plants).

This manager and others felt that the Arab countries never got over their suspicion of why SCP was stopped in Italy and Japan. They felt that if SCP was not good enough for Italians to eat, it certainly was not good enough for Arabs – this was especially important since they were unable to judge the technology for themselves, and were therefore looking for independent evidence of the value of SCP plant. Several managers involved in licensing negotiations thought it unlikely that they ever understood what really happened in Italy and Japan, but that it was enough that something unspecified had gone wrong and money invested in SCP had been lost.

It is the case that when oil prices dropped in 1986 the Arab countries returned to look at SCP once more. One of the former BP consultants declared that he had been asked to prepare reports for Algeria, Saudi Arabia and Kuwait in the middle 1980s, because of the oil price fall. Nothing has so far come of these reviews – SCP remains a technology-in-waiting for a different world economic price relation between hydrocarbons and food.

CONCLUSIONS

The ideas of 'use' and 'need' in the market concept

This chapter traced the origin and development of market ideas in the SCP projects from early in the innovation process. It supports the idea that innovation involves interaction between both market and technical concepts throughout the development process. It is argued here that this is best understood as a cognitive process on the part of the managers.

The origin of the innovatory projects was a crude 'match' between undeveloped, untailored market concept and existing technological ability that could feasibly be deployed to meet the market idea; this vision of a potential technology–market match was necessary to set the project moving. The crude match was then elaborated by giving physical form to a prototype technology and developing the detail of the market idea. Many SCP projects began because of the appealing nature of the match between some form of adapted fermentation technology and a looming demand for 'food' world-wide, but in the detail of the elaboration of this match some projects had markets while others had not.

At the early stage of the projects the crude 'market idea' does not necessarily differentiate between an intended 'use' for the new product and whether there was what we might call a real market 'need', where people would pay for the use of the product. Even SugarCo's hole-in-the ground fermentation technology had what might be called a 'use-concept' behind its development, but since SugarCo found no one ready to pay for its use, it could not be said to have a market. The core issue for the management of the project is whether the intended 'use' of the developing technology fulfils a 'need' that will bring revenue streams to more than cover the cost of development. While it is all too easy to build 'technologies with uses', it is just as easy to confuse use with need, especially if the firm has no experience of the target markets. This judgement then rests on how information outside of the firm is gathered and interpreted and is typically the job of the marketing function in large firms.

Two types of market concept

For those firms that developed their projects up to the commercial stage I suggest that it is useful to define both an *innovation market* concept and a *reference market* concept. The innovation market concept is the projected market for the innovatory product, the mental construction of 'how the product will be valued' that managers use to guide the development of the new production technology. The reference market, on the other hand, is an existing market based on real, traded products, which is understood in most detail by those producing and those consuming those products.

Both market concepts can be broken down into sets of qualities which are used to gauge the value of the traded product in the case of the reference market, and the innovatory product in the case of the innovation market concept. The split between the two types of market concept is justified because the managers of the innovating firm largely construct the innovation market concept from the reference market. The reference market is so-called because it is the market conception on which the innovating firm managers draw most heavily in the process of constructing the innovation market concept. This construction of the innovation market concept is a cognitive process of understanding the qualities that make up the reference market and selecting those that are valued, for inclusion in the innovation market concept. This process may not be taken very far or it may be very elaborate – but upon it depends the degree to which the new production technology will have an adequate market understanding built into its hardware and operating practices.

It is not only the reference market that is drawn upon to build the innovation market concept. Indeed, the usefulness of the reference market is limited because it is itself a mental construct built to articulate qualities that usefully differentiate between company products *already traded*. If these qualities do not differentiate between existing products they may not be articulated as 'formal knowledge', that is, as knowledge transferable in language or written form.

Hence the example of Pruteen's selenium deficiency, where minimum selenium content had not been a criterion for judging animal feed quality in the reference market, that is, among the feed compounders, since all compounded feeds that included fish-meal contained sufficient selenium. The idea of specifying minimum selenium content was only relevant to microbial feeds as a novel source of animal feed, because they were deficient in selenium. This element of the innovation market concept was formally articulated by ICI's R & D department – that selenium should be added to the innovatory feed to reach a minimum selenium level – and became part of the innovation market concept.

The oil crises threw BP and ICI into the position of having commercial-scale plant and sunk investment without good market outlets. The attempts to licence the technology and, in ICI's case, the investigation of the fractionation project were further attempts to match technology to market ideas, only this time SCP technology had been brought into existence. As if SCP technology was an overextended limb of the firm's base technology, it was eventually diagnosed as not worth any further development effort and excised.

It is too trite to classify these projects as 'market failures', since, as I have tried to argue, a 'market idea' is almost always present in projects because they almost always have an intended use. The issue for the innovating firm is the blurred nature of the distinction between use and need and the difficulty in discriminating between the two as they have been defined here. They do so by building an innovation market concept, perhaps from previous experience of the reference market or through acquiring that experience for the first time, and are effectively developing criteria by which to judge the developing technology as being merely useful, or having a use which is needed. The quality of the internal debate over the definition of the market concepts and how they should guide the development of the new production technology is likely to influence the success or failure of the innovation.

4 Cooperation and competition between companies

While the last chapter looked at the environment external to the firms and their projects, this one examines inter-firm relationships. The development of SCP was sufficiently complex that companies could compete and cooperate at the same time but on different aspects of SCP development. In general the producer companies competed on selected project development characteristics, but cooperated on regulatory issues and promoting the image of SCP generally. Since none of the products were traded in quantity, the selection of which characteristics were commercially sensitive and which were not was influenced by experience, and expectations of the form of the mature SCP industry.

The SCP companies were all attempting to obtain a share of the animal feed market, so the real competitor product was the standard fish-meal/soya-meal animal feed and the competitors were the soya growers. The innovative SCP feeds were intended to compete on quality characteristics and price differences thought to be obtainable because of their radically different production technology. This was meant to be raw technological competition, the threatened (partial) replacement of one system of production technology with another.

With a world market for soya-meal of some 13 million t/a, output from a couple of 100,000 t/a SCP plants would be negligible compared to the size of the total soya market. Even if the first SCP plants had proved successful, it would have taken years to build production to levels which had an impact on the markets for soyameal. So SCP companies were not an immediate threat to soya consumption. The enormous size of the bulk soya-meal market also meant that the time when prospective SCP producers would compete for market share was so far off that it reduced producer rivalry. A project manager within Shell accepted this view and believed that as a result the cooperation between company research teams was greater than was usual for a commercial project. However, there was competition of a kind and this manager did see the SCP companies' relationships as technically competitive, commenting, 'Competition? Yes, we were very aware of what they [other companies], were doing, mainly for what they were claiming for their SCP products . . . colour, texture and so on. Yes, minor characteristics'. This variety in SCP products was the subject of intense interest by the producer companies for the clues it provided to other companies' process designs, knowledge of which would inform their own development process. This was competition a step back from the market, competition that worked on the basis that you should not give anything away for nothing, even if you could not see how it would hurt you in the short term.

BP AND ICI

For BP the judgement that a company was a competitor depended on how similar their process was to BP's and how close in time they were along the development path. The Japanese were the real competitors to BP because they had planned to use the alkane route to SCP and at the same time as BP. In addition the Japanese companies were seen as potentially powerful competitors because they had fermentation technology skills that BP lacked. BP began talks with the Japanese which were intended to tie them up to some type of deal; one suggestion was for an exchange of fermentation skills for R & D skills, which would have ensured that BP was not disadvantaged in the new industry. BP licensed its technology to the company Kiro Hako for a sum which one manager claimed paid for BP's research and pilot plant phases. With the end of the Japanese companies' projects in 1973, BP had a clear field. ICI was not a competitor because they started development so long after BP:

> They were coming along, snapping at our heels, but I don't remember seeing them as a challenge, no. If we had had no problems we would have been producing in 76/77 and we'd have been perfectly able to tolerate ICI going into the same market as well . . . they were not a rival but a younger brother to be patted on the head.

One of the BP managers described the ICI process as 'second wave' technology which would be on the market when the BP process was at its peak, giving BP time to develop a third wave of SCP technology ahead of ICI. So there was the idea that because the technology was likely to continue to change, the immediate intensity of competition was weakened. However, one of the ICI managers commented that in at least one of the early ICI meetings with BP, BP did not realise how committed ICI were to SCP and tried to warn ICI off a technology which 'belonged' to BP. Once they realised that ICI was serious about building a large-scale plant, they accepted them and even worked with them on some common problems.

BP was seen as the pioneer in SCP by ICI and by other oil companies. While the Agricultural Division in ICI had its own reasons for developing SCP, many oil companies began research largely because they saw that BP believed in SCP and its potential profitability. They were unwilling to allow a traditional rival to develop a major new technology based on hydrocarbons and to which they had no expertise and no access.

This was the principal reason for the Institut Français du Pétrole (IFP), ELF and TOTAL involvement in SCP and one of the reasons why Shell and Esso became involved in the technology. The IFP maintained a programme of research on SCP from 1966 until the late 1970s, with which the French oil companies ELF (who joined in 1968) and TOTAL participated with the aim of keeping abreast of SCP developments at low cost. If Toprina succeeded they wanted to be able to follow BP into the SCP business quickly. ELF contributed research personnel and money while TOTAL only paid a share of the costs and followed the work through seconding one engineer. The three partners also created an economic interest group[1] called the Groupe en France de Protéines which continued until 1976, when CFR and Total abandoned the project.

While we might say BP and ICI were weakly competing, they did not consider themselves to be competing with RHM at all, since that company was trying to reach the human food market rather than the animal feed market. Nor were the village technology projects (see Appendix) seen as competition, since both the market and technology were sufficiently different.

PATENTS, SCIENTIFIC RESEARCH AND COMPETITION

ICI attempted to block new entrants to SCP by

> (a) good patenting and (b) control of the raw material. ICI has licensed or built 70 or 80 per cent of all the methanol plant in the world . . . we could control methanol production by the catalysts essential to produce it from methane, if not by the licence for the plant. All methanol plant uses ICI licensed catalysts.

The French research work on methanol processes never advanced beyond laboratory-scale work and when asked why, one ICI manager thought the ICI ownership of methanol patents to be a great disincentive to the French:

> They were late coming in and when they did they found BP and Liquichimica were well established . . . and second, ICI was going down another path and presumably had the patent position tidied up without any gaps.

BP and ICI felt they had the advantage of being leaders in their versions of SCP technology and that they were successfully deterring new entrants through control of the technology.

When a company registers patents on a new technology they make some of their knowledge public while retaining monopoly rights over its exploitation for a set period. In the period up until the company patented its research, it tries to keep its activities secret. One of the RHM managers described the company attitude:

> there is a natural inclination not to share information which has taken five or six years to assemble, especially regarding the organism. We were obviously much more relaxed when we had established a lead by way of a body of data . . . and had patented the organism and the process that surrounded its growth.

Companies were certainly not interested in academic publishing, although many of their research scientists wished that they could publish. One of the SugarCo research managers commented on the persistence of the desire to publish among some of their research people,

> who, generally speaking, don't like industrial research because they feel they will do all this work and someone will lock it away in a filing cabinet and they will never be allowed to use it.

Companies would compromise and allow their scientists to publish in order to attract the people they wanted, but the publications would be scrutinised for commercial disclosures. At ICI this meant that a prospective paper had to be signed by four different groups of people; the research manager, the research director, the business area manager and possibly even a Board member's signature were all required before a paper could be presented at a conference. Even talks, with no written proceedings, would be checked by business managers. One ICI researcher found that he learnt what was permissible through these formal checks then applied the same checks in informal discussions with colleagues, and he commented that, 'Sometimes it's ludicrous, and they nit-pick like hell . . . but the business is what pays the money for the research, so we can't upset the business'. The result of this elaborate checking for disclosures was, as a BP research manager pointed out, that you learnt nothing of commercial significance from scientific papers: 'They are as bland as you care to make them. No company worth its salt publishes anything'.

If the formal papers were innocuous, informal and personal contact was another danger area for companies. Scientists were supposed to develop a code of personal behaviour with regard to rivals at conferences and when talking with former colleagues. One of the ICI researchers described how this code was acquired and applied at a conference:

> you're left to your own judgement how much to tell them [other scientists] . . . nobody ever teaches you, you find out from experience what is and isn't sensitive . . . and some will think that it is, and some that it isn't . . . get someone buying a few extra pints on the night of a conference and there is no knowing how much some people will tell you . . . you've got to be careful you don't go wide of the mark and by discussing the problem you don't actually give away the process or something like that.

In most cases it was easy for a scientist to know what was commercially sensitive. Basic science knowledge was exchanged freely, such as how organisms grew in general and how the energy flows at different stages in their biochemical pathways, because these were of interest to all the scientists and helped all companies. On the other hand, talk about how to grow organisms at high process efficiency was commercially sensitive because this was the characteristic by which companies were selecting one organism over another and it related to the commercial necessity of attaining economies of scale.

Commercially important information did leak out and using logic and intelligent guesswork much could be deduced of a company's commercial targets. So one research manager commented that

> Over the table in an unguarded moment, people say things they shouldn't say. And people move from one company to another, that's how it happens. Companies don't publish, the most secret things they don't even patent. The last thing you do.

Different issues were sensitive to different degrees, and one former SugarCo fermentation consultant gave what he thought was the hierarchy of commercially sensitive topics:

> The most important thing was always the organism. Number one. Absolutely. You don't actually give it a name, or you keep the name secret. If you look down the microscope . . . well, a yeast is a yeast. So it's the cultures and where they keep them. The second would be the medium, not the single medium, but its make-up.

The companies would admit that the medium was methanol and so on, but the precise concentrations of minerals required to maximise growth would be kept secret.

> The third thing would be the type of process; there is quite a bit of know-how there, growth rates, balance of amino acids, (percentage mass that is protein). The fourth thing would be separation, that is a very key area indeed. In the Pruteen process a lot of technology was developed to separate the organism. The final thing is what you do to it once you've separated it, downstream, that is a key area too.

To a degree this list varied depending on the company, while it also changed through time. So RHM referred to their organisms by numbers, C1 and A3/5, but BP named their yeast as a strain of Candida Tropicalis. The same former fermentation consultant thought that RHM eventually named their organism because it became clear that unless a competitor had the exact strain, they could not reproduce the same results, so the name was of no use to anybody. He recalled the release of the name of RHM's first organism:

> RHM kept their bug secret for a very long time . . . everyone knew it was a filamentous fungus and we had a very good idea what it was. I remember it leaking out, it was a Fusarium you know, and then everyone wanted to know what sort of Fusarium, and when they discovered that the first organism was a Seletectum strain they all went rushing round looking for Seletectum strains. But they didn't get the right one. And then lo and behold, RHM chose a different strain anyway.

Certain areas became less secretive as time passed because people realised that the skills of fermentation were extremely difficult to acquire, even with the name of an organism and even with its optimum nutrition requirements, as one BP manager commented:

> Monoseptic continuous fermentation is the sort of process where you can give someone the name and pedigree of the yeast, a

sample, but if you have not got all the 10 to 15 years of culture experience, medium optimisation, its hardly worth the effort. So you can publish lots of facts and figures, they don't help you to ferment. Fermentation is an art and a skill, you don't pick that up by reading about it.

On the other hand certain events had resulted in a general increase in sensitivity to commercial matters among the microbiological community. The story of the public release of the discovery of monoclonal antibodies now represented a 'cautionary tale' among microbiologists:

In those days (1970s) academics were extremely free with their ideas in the UK, and we've seen some of the consequences of that too. Milstein and monoclonal antibodies . . . it was at a time when academics would go along to a scientific conference and talk about their work and not appreciate the commercial relevance of it . . . that has changed.

RHM: COOPERATION FOR FINANCE

RHM differed from the other developers in that they felt they had insufficient resources to research and develop their product as they wished. This led them into a succession of deals with companies where the second company would support the project financially in return for rights over eventual profits: 'At that time we were not a rich company and we didn't have the money to carry this one through on our own account'. RHM searched for a suitable company and came up with Du Pont. Du Pont subsequently supplied money and some engineering input to the project for three to four years, until a period of retrenchment in Du Pont when the joint project with RHM was cut. In the mid-1970s RHM worked on product development with Unilever for a time, where Unilever experimented with myco-protein by treating it as a raw material, putting it in pies and testing consumer reaction and giving feedback to RHM. Then in the early 1980s the British Technology Group was approached and they agreed to support the project financially, during a period when RHM profits were under pressure and it would normally have been impossible to justify such a high-risk project.

Finally a relationship with ICI was established that continued into the 1990s. ICI had first been approached in the late 1970s and an RHM manager recalled their reaction:

they said, you have a very nice little project here. We are working on lots of problems that you are working on. We have to tell you that (a) we don't know how you did it for the money because it cost us 40 times as much, (b) we need another protein project like a hole in the head. We have enough problems with our own, thank you very much.

But once ICI lost its commitment to Pruteen they were ready to discuss the possibility of collaboration. In the early 1980s RHM and ICI managers were meeting to share views about obtaining regulatory clearance and found that they had many complementary interests. Meetings to explore the chances of cooperation followed these first contacts and the two companies found a way of working together. ICI and RHM managers described the rationale for cooperation in similar terms, with one commenting that, 'the ICI relationship is a combination of sharing risk, costs, skills . . . about bringing together a whole complex package of, I think, complementary skills and interests rather than competitive skills and interests'. ICI was international and had the ability to design and build large plant. RHM had the ability to develop complex food systems, a number of market entrées in Europe and food marketing expertise. Production of mycoprotein was begun on ICI's Pruteen pilot plant, which was standing idle, while a jointly owned company called Marlow Foods was given market development responsibilities at High Wycombe. Research on mycoprotein was split between ICI's Biological Products Division at Billingham and RHM's High Wycombe food laboratories.

Marlow Foods also worked especially closely with Sainsbury's to market the first mycoprotein products. Sainsbury's could offer Marlow Foods access to 16 per cent of the UK shelf space for meat, they owned their own meat pie company and could offer RHM knowledge of the meat products' market and meat manufacturing expertise that RHM did not possess.

COLLABORATION IN DEVISING A REGULATORY REGIME

While certain details of the processes were guarded, in the areas of regulation and public perception of the product SCP developers had reasons to cooperate. Research scientists recognised that it would help legitimise SCP as a safe and respectable industrial product if they referred to each other's work whenever they delivered a paper at a conference; so researchers from RHM would mention Pruteen at food industry conferences to which ICI scientists would not normally go, while ICI would refer to RHM's work at chemical industry conferences.

BP and ICI were anxious that the new regulatory guidelines that had to be drawn up for SCP would be as favourable as possible to their SCP products and as unhelpful as possible to potential new entrant companies. To this end the two companies worked together in approaching UNPAG[2] and the food safety authorities in the UK and the EC. The aim was to agree the kind of regulatory framework that would best suit the two companies' SCP projects and then to present the same arguments to the authorities when they began drawing up a framework.

An ICI manager described BP's role in obtaining guidelines from UNPAG as enormous. This organisation was set up in the 1960s in response to the protein gap and was concerned with ways in which the UN could overcome world famine through promoting the development of new protein sources. It was important to BP that it picked SCP as one of the solutions to this protein gap, and this eventually happened: 'BP saw an opportunity to get the UN thinking about SCP . . . and from about this date, 1973, there were several conferences, I think one a year, until 1978'. This manager described the type of regulatory regime that ICI and BP favoured:

> Individual projects had to be tested rather than have generical tests for SCP. The EEC directive on SCP reflects UNPAG thinking in this way too . . . our approach to testing was conditioned by our need to deter late entrants. Because if each product belonging to each firm had to be individually tested it would cost between £2 million and £3 million, at 1976 prices, for each company. This in fact became a very marked disincentive to the late new entrants because by 1976 specific, non-generic guidelines had been accepted by UNPAG.

In the case of the EEC regulations, ICI were only becoming interested in influencing the regulations when BP had already decided to close their Sardinian plant. ICI had to persuade BP to work with them, even though there was no longer a gain for BP – but nor was there much of a cost either.

BP had a special reason for wanting individual product, and not generic testing procedures from the EC. Liquichimica's SCP was produced by licensed Kanegafuchi technology, which had been attacked for having had an inadequate amount of animal testing research in Japan. BP were arguing for specific testing procedures with the Italian health authorities while Liquichimica were arguing that their process was a yeast grown on alkanes exactly like BP's and so if approval was given to BP, it should also be given to Liquichimica. At the time of the debate it would have been of great value to BP to have had the non-generic EC directive on SCP testing procedures, but by the time these emerged in 1984, after eight years of debate, they were of course useless to ICI as well as to BP. Managers in both BP and ICI thought that the Italian authorities could not approve the BP process while refusing to license an Italian company's process; so we have another rationale for why the Italian authorities preferred to stall and not approve any process at all.

However, the late 1980s venture-capital-based Dansk Bioprotein believed that they would benefit from the EC regulations, if only because there were procedures where before there was nothing.

CONCLUSIONS

SCP competed with orthodox animal feeds on the basis of its radically different production technology; it threatened to capture market share through technological competition. No SCP products from the purpose-built plant were widely and continuously traded but because the producer companies expected to be trading they did see each other as potential rivals, although the strength of the perception was linked to the similarity in target market and process technology – the 'generational' difference between the BP and ICI technologies weakened the sense that they were in competition.

There was technical rivalry and an evolving awareness of which technical process characteristics were critical if a rival or new industry entrant wanted to copy the process. To an extent companies released more information as time passed as they realised that the skill of fermentation was difficult to acquire except through years of research and experimentation, and as they registered patents on their process designs. The SCP producers had relationships which were a combination of competitive and cooperative elements; they tended to compete over individual process characteristics which might differentiate between their products and tended to collaborate where that would establish SCP as a generic product – on the exchange of basic science the need to establish a regulatory regime and a public perception of SCP as a respectable industrial product.

This shows that competition was not an entirely straightforward issue. Companies both drew on past experience and learnt in the process of development which characteristics of the new process technology were critical in terms of future competition. The pattern of competitive and cooperative links was maintained through deliberate management, whether, in the case of competition, it was through controls on research scientists or whether, in the case of cooperation, alliances had to be built through a process of negotiation and sometimes persuasion.

This view of competition, as something which is linked to particular process and product characteristics and which may coexist with cooperation, can be contrasted with the simple idea that competition is both uncomplicated and that the more there is the better.

5 Firm and industry culture and the role of senior management

In this chapter we look at managers' perceptions of how firm characteristics influenced the development of SCP projects. This includes issues of whether a firm or industry 'style' or 'culture' exists in a firms' approach to project development, and it is the style of the firm as observed by middle-ranking managers that dominates the first sections of the chapter – later, structural features and the role of senior levels of management in the projects are considered.

CULTURE AND STYLE IN PROJECTS: DIFFERENCES BETWEEN RESEARCH STYLES

Many of the novel food fermentation innovations required a new combination of working relationships between scientists and engineers from traditionally separate backgrounds, often micro-biologists and chemical engineers. The multi-skill development teams and the planning of the construction of fermentation-based plant in companies that had no previous experience of fermentation led some to identify both differences in research style between companies and a general approach to innovation typical of the large companies.

One of the research scientists working on the mycoprotein project felt that where there had been friction between the two companies' styles it was at the researcher level:

> the meeting of cultures has occurred at my level . . . everything at senior level has been OK, so in terms of companies like mammoths colliding, it's not on, because it's been the troops who have been fighting it out; the generals have been having cocktails together.

The same researcher contrasted ICI and RHM research:

most of the ICI people would probably say that fully understanding things is the way to go before you can really build the business, whereas some of the elements in RHM say we understand as much as we need to . . . if it works it works.

The ICI approach was characterised as one which sought to understand as opposed to an empirical or pragmatic style. The cost of the empirical style was that when a process did not work, it would not be understood, and simple changes to the process that could correct the problem might be missed. When the two companies came together in the mycoprotein project, they moved the process research to Biological Products at Billingham and the recipe formulation stayed with RHM, and one could say that each company controlled that part of the mycoprotein project that suited its style.

> a lot of the things we inherited in the process were purely empirical . . . people of my background started coming in and asking, why do you do this, and why that? Nobody could answer, so my role in the first year was to challenge dogma or myth or whatever, accepted wisdom . . . find what was right and what wasn't.

One researcher suggested the difference in styles was financial in origin. Mycoprotein research at RHM had always been short of money, compared to Pruteen, in part because of the relative sizes of the companies. Less money meant there was less time and money to spend on the 'understanding' of the various mycoprotein processes.

Within RHM, one manager agreed that ICI were more used to applying rigorous fundamental science to their problems than RHM, but if RHM was compared to the rest of the food industry, it stood out as unusual because they did apply a more rigorous science approach to some of the relationships that underpinned food – most of the food industry remained a low-technology, low-science industry.

There were other general differences between the companies that managers noticed, based on the different industries of the two companies:

> ICI's posture is somewhat more aggressive than ours. They are more aggressive world-wide . . . they are prepared to test by pushing what the boundaries are. RHM has more of a partnership style as they are very, very, very sensitive about anything that might damage their reputation for food.

ICI's products' performance were both more independent one from another and from the company's reputation, RHM's reputation would be endangered by any one product failure and RHM depended on its good company image to continue to sell its other products. So in the mycoprotein project, RHM was given responsibility for marketing and a slow, careful development of the market became the policy for mycoprotein.

A chemical engineering culture

ICI's chemical business required the employment of many chemical engineers, and chemical engineering was the dominant background of ICI managers. Some managers believed ICI had a progressive record in creating interdisciplinary teams to research and develop particular projects, but many inside the company pointed to the predominance of chemical engineers as something which was not wholly beneficial on the Pruteen project. One referred to what might be called a chemical engineering culture:

> here the chemical engineer was king, because ICI always used to recruit the bright young chemical engineers from Cambridge and it used to be a self-evolving system . . . now, in Biological Products there are still quite a few chemical engineers about but their status and their influence are waning slightly.

The success of the chemical engineers in creating and supporting ICI's core businesses was perceived by some to have led to an arrogance when it came to considering alternative approaches. A research manager in another company remarked how a microbiologist friend of his working in ICI

> was always saying, you have no idea the grief I get from a particular divisional director, because he doesn't see why I'm messing about with bacteria when what he knows about is pumping ammonia around reactors and catalysts. You are doing something which the body of the organisation doesn't understand. It's viewed with suspicion.

There were differences in approach that had to be resolved on the mycoprotein project:

> An engineer's criteria are throughput and yield, and so on. The fact that he might have altered his process to improve the yield, which knackers the quality or whatever, in a sense doesn't really matter to them. This is being a bit callous about it all, but engineers, if you knock the quality down by 10 per cent and

improve the throughput by 10 per cent, they say, well, let's sell it for 10 per cent cheaper or something. Then you say, you just can't do that, its either good or it's not good. There is no sliding scale. People either like the food, or they don't.

One of the project managers on the Shell SCP project believed the engineering background of senior management at Shell created difficulties for biologically based technology, like SCP. There was a tendency to apply rules of thumb to judge biotechnology projects that originated in their engineering experience.

If you are a chemical engineer and you are given the bones of a process, you will probably accept that if you work hard enough you might be able to improve the yield by twofold or maybe fivefold, but that is probably the limit due to the nature of the chemistry. But in biological sciences that is not the case. You know perfectly well that once you have the biological reaction going, you are certainly going to increase the yield tenfold, with a bit of work a hundred-fold and a thousand-fold is down the tracks . . . it's very difficult if you are doing biological R & D to convince someone who is not a biologist that this is actually true. . . . So they tell you, come back when you've improved it, but to do it requires quite a high level of resource . . . very difficult to put that message through to the kind of culture we had in Shell at that time.

A Shell research scientist remarked that 'You tend to draw battle lines, some people are very divided on it, especially SCP . . . the chemists have an enormous prejudice against biotechnology'. Another former Shell SCP researcher has written a paper on chemical engineering and biotechnology which supports the idea that there is this split in views based on training: 'two important orientations exist, i.e. the approach based on biochemical engineering and the approach based on microbial physiology' (Hamer 1985: 346). Hamer goes on to claim that the biochemical engineering approach has largely failed over the last 25 years, despite its being favoured by educational institutions when training biotechnologists.

ICI's first food scientist had found working in ICI frustrating to start with: 'I must admit, I found this bloody annoying at the start, why should I have to convince a chemical engineer that what I knew about food science was correct?' The process of establishing his credibility took him almost a year and he epitomised the kind of attitude he met during this year, especially amongst more senior chemical engineers:

no one's being aggressive or anything like that, but the view is almost that the food scientist is someone who stands in the corner and stirs the pan with a wooden spoon. An awful lot of people thought, 'a food scientist, you must be good at cooking then?'

From a food scientist's perspective, the attempt to modify the Pruteen process to make a food human beings might eat was not something you would do; you should start with a product that people would eat, and then work back into the process design.

Big is beautiful

Many research scientists and managers referred to the size of companies to explain their behaviour or potential behaviour with respect to SCP. So a research manager from SugarCo commented on ICI:

> they couldn't look at the sort of small market we could look at. Almost as if the size itself imposed the requirement that, hell, we're not interested in anything that is less than 100,000 tonnes per annum, because that is what we know. There is a house style, and in ICI it is part imposed by size . . . the kind of research we did in SugarCo would never have started in ICI.

One of the ICI SCP managers thought the culture was more 'if it is big it is better' and that, although it had waned in recent years, it had been the correct approach to SCP, since SCP plant had to be big to reap the economies of scale necessary to be viable. A project manager in Shell thought that certain ways of developing SCP technology were closed to companies like Shell simply because of their style and their size, commenting,

> a company like Shell, they couldn't think in terms of human food, it was difficult enough to think in terms of chemicals. Shell are in chemicals . . . [but] the concepts that define the chemicals business, selling things in Kg, its just too small for their operation.

Du Pont, with a similar approach to ICI, changed the direction of the RHM project for a time,

> You have to say the vision was lost a little bit around about the time of the Du Pont relationship, a combination of growing oil prices and seeing the complexity of the task ahead, people started to turn to some kind of dried form of the product, a burger extender, particularly in the Third World. Du Pont were very keen

on this because their idea of a world-scale business was, how can we make it as cheaply as possible, how can we sell it as widely as possible? What is the easiest form to go and sell the product in?

Du Pont and RHM were sufficiently convinced that this was a better approach that they began a joint project with the Iranian government in the mid-1970s, to build a plant to provide a biscuit protein supplement for school children, a project which was ended by the Iranian revolution.

ROLE OF STRATEGY AND SENIOR MANAGEMENT IN SUGARCO

The previous two sections examined the role of style and culture in the projects, but these are characteristics of the firm which are not primarily linked to formal structures within the firm. The following sections move towards the more formal and deliberate choices made by the firm, the area of strategy and structure and the role of senior management and the board. Crucial to this chapter are the relationship of the R & D department to higher management and how and why projects were selected for support to different stages of development.

In general senior management was important when a project began to demand finance that was an important fraction of a company's total resource; the decisions to build the full-scale plants of BP, Liquichimica and ICI required the involvement of the companies' most senior levels.

A strategy of diversification through the R & D department

The SCP projects investigated by SugarCo were a consequence of a restructuring of the company and a linked decision to increase the resources available to the R & D department. During the restructuring the then head of R & D, Finlay,[1] persuaded the SugarCo board to accept his idea to use the R & D department to spearhead SugarCo's diversification away from sugar; until this time, research in SugarCo was restricted to improving the sugar-refining process.

In the mid-1960s the SugarCo board seized an opportunity to buy a molasses trading company on very good terms, but the company had as large a turnover as SugarCo and it became obvious to the board that assimilating this company into the existing structure of SugarCo was problematic. The SugarCo board asked the consulting firm McKinsey to review the company structure, and McKinsey duly suggested that SugarCo adopt a divisionalised structure. This involved a major shift of operating responsibility from the board to what was called EXCO, the Executive Management Committee, which was made up of the heads of the new divisions and included the head of research, Finlay. The committee met as often as it felt was necessary to operate the day-to-day business of the company. This was something that the board had supposedly been in charge of under the previous structure, but according to one manager,

> There was no one on the main board . . . and until EXCO was created and the head of research was put there, there was no one at any senior level who really understood what research was doing . . . and the main board met something like once a month, except in August, so basically they couldn't run a paper bag factory. EXCO met, if necessary, on a daily basis, they met on a functional requirement basis.

The change to a divisionalised structure was complete by 1970–1 and for this research manager the effect on R & D of the creation of EXCO and its inclusion of Finlay was that

> it meant more money . . . we had a pipeline to the top, we didn't have to convince anybody except Finlay. When I wanted to build the [SCP] pilot plant I was talking about a lot of money, because we had never done this before. If I could persuade the head of research, he would make the case to EXCO.

He was sure that without this reorganisation a pilot plant for SCP would never have been built because the board would have been responsible for taking the decision to proceed. The SCP project manager felt that at no time was the board capable of approving such investments or of judging research activity, because

> the board had not got a clue what was really going on. . . . How far did senior management's ignorance of technological matters affect its decision-making? Totally. Totally. The successful research-orientated companies usually have an ex-research man on the board.

The SugarCo board did not have this kind of R & D expertise available to them. Directors had responsibility for particular divisions of the company sand tended to represent 'their' division's views to the board. This project manager described board practices which further reduced its ability to intervene knowledgeably in research matters:

Because it was a family company the main board was essentially [family members] and what we used to call the 'hangers on', who had married into the family. . . . In those days there was not a thought that this director ought to be responsible for R & D because, damn it all, he knows more about it. It was not a functional appointment, it was more or less, you've had it for three years Col, now its my turn, sort of thing. Because R & D is a toy that lots of directors of large companies like to play with, particularly when it's successful.

Within a year of the creation of EXCO in the early 1970s, the board decided to use R & D to spearhead diversification away from sugar. A research manager described the decision as,

a consensus, I don't know who championed it. I think Finlay himself put the idea into the research director's head, again through his technique of talking into the wee small hours. He had talked to other members of the board and they had thought, generally speaking, this is a good idea. I suspect the decision came from below rather than from above.

The R & D function was now reorganised, away from the sugar refinery servicing function it had once had. A new direction for research was worked out, based on the raw materials of the business; sugar, molasses and bagasse (the dry pressings from cane sugar production). The head of fermentation research described the results:

What you finish up with if you think it through, is a matrix; we used to do this . . . in management meetings. They plot what you are good at, your functions, by your research needs. So you say, what happens if we apply our skills in fermentation and bio-chemistry to the need, some other use for sugar? Out of that comes an SCP programme. But you also have to look at bagasse and molasses. And again, as regards this skill, you end up with SCP, to produce alcohol and liquid waste disposal. . . . You have a research scheme, not original, but very effectively used by Finlay. He said, I want four basic departments for the basic skills, these are biochemistry, organic chemistry, fermentation, and physical and inorganic chemistry.

The decision to rely on R & D for the diversification away from the traditional sugar-as-sweetener business occurred because the head of research was such a strong champion for the idea. With the concern over the future of the business if it continued to rely on direct sugar sales as background, he persuaded the director of research, who convinced the board, that this was the right course to take, and the board were, perhaps, unusually open to suggestions that would appear to solve their problem. The detail of what the R & D department's expanded activities would be was left entirely to the head of research and those he chose to include in the process of creating a research strategy.

SugarCo's decision to build and site an SCP pilot plant

Finlay already had an interest in SCP the carob project being researched by two technicians in Greece. He created the fermentation group within R & D and brought in someone to run the group whom he trusted to do the job effectively, who was given to understand that the carob project would have to continue as part of the group's activities. The head of fermentation had the management problem that his scientific staff in England resented the enhanced status of the two technicians who were working on the project in Greece. Although they were only qualified as technicians, they were effectively running the Greek project, something that would not be allowed in England. He solved this problem by expanding the SCP project so that the Greek activity became only a part of the whole, part of a project to test a variety of different substrates for fermentation suitability. In this way the solution of a management problem helped create the SCP activity, although the head of fermentation was interested in SCP in its own right.

In the mid-1970s there appeared to be a chance that the Cyprus government would be interested in acquiring the carob fermentation technology. The head of fermentation research persuaded Finlay that a pilot plant was necessary to continue research,

> Scientifically, we needed a pilot plant somewhere under third world conditions . . . so often when you are running a project it is a personal thing. I thought, I had done a spell in Thailand, seconded to the UN development programme for a while. I said, I like Thailand, we should consider putting a pilot plant there. But from board level it was made known to us that if we were going to do this it would have to be in Belize.

The head of fermentation described how the economy of Belize is dominated by the Belize Sugar Corporation, itself effectively controlled and managed by SugarCo. Such companies as SugarCo were viewed with deep suspicion by liberating governments who knew they had to have them, because if no sugar was exported they would have no income. So companies like SugarCo come up with what I call the political project. The Belize plant was one of them.

The head of fermentation thought Belize was a particularly poor choice for a plant.

I said, for God's sake, why Belize? I put up all the standard arguments. To start with, there is no agricultural waste – 'Don't worry about that, the plant has to be in Belize, the government has requested it, and we have decided that that is where it is going to be.' So the usual fanfare of trumpets and the PM of Belize came and we all stood around looking suitably enthusiastic. I was left with the problem, what do we do? Because really there is nothing there . . . which is why we worked on citrus waste, because there is a citrus canning industry and it uses sugar, so there is citrus waste . . . which is fed to animals directly – and they seem to like it . . . the last thing one would choose, citrus waste, very special characteristics. Highly acid. A so and so of a time getting the thing to work.

The plant had never been an economic proposition and so, 'After two or three years, when the company judged that the time was ripe, they quietly shut the thing down and stole away . . . that's a fairly normal story for a company like SugarCo'.

SugarCo board's control of research

The research department embarked on an exceptionally free-ranging programme of research by the standards of other industrial R & D departments. The matrix by which the various projects were grouped was a guide for the application of technical skills to the raw materials of the company and was intended to broaden the range of projects, an aim which it achieved admirably. Provided that the fermentation department could think of a technically feasible project connected to sugar, molasses or bagasse, it had a good chance of being investigated. Unfortunately there was no formal marketing input; no marketing department involvement and no means of independently judging the market potential of the projects. It was left to the R & D department to judge their commercial potential. A research manager described how the board would enquire into their reasoning on SCP and how they would reply:

> Our decisions . . . I think this is important . . . that the decision
> to go into SCP never came from the board, it came from the
> bottom upwards. We indicated to the board . . . they'd say, why
> are you doing this, and we'd say well, because we think it was
> successful in this and may lead to that and that. That often
> happens, decisions don't come down from the board to research,
> in those years it came up the other way.

Through the 1970s the board did not apply its own selection criteria
to particular projects like SCP; they were content to devolve
responsibility to those within the R & D department, asking only for
project information and showing no sign of disagreeing with the
judgements of senior people in R & D.

By 1976–7 a new division was set up to manage the development
of the R & D projects which had commercial potential. This was
called 'New Ventures' and the head of fermentation research was
promoted to manage the division. The new head of fermentation had
joined SugarCo from Glaxo and not surprisingly had research interests
in a different area. Once in charge of fermentation he began the
process of closing the SCP projects and in his opinion

> it [SCP] was not something I could put forward to the board for
> assessment . . . just didn't make sense. For SugarCo, proceeding
> with the project was an R & D decision and the closure of the
> project was an R & D decision . . . perhaps one might question
> why there wasn't closer questioning from the board given the
> amount of money that was spent on it . . . but it's probably a good
> thing in the R & D world that you can hide and get things done
> . . . if you believe in something you can proceed with it without
> it appearing to be a major item.

His view of SCP was that

> it was a typical example of the dangers of research push, it was
> very easy, early biotechnology. Good public image, popular, the
> board of directors would understand that sort of thing in very
> simple terms; muck-into-protein, feed-the-starving-millions.

The board was really irrelevant to decisions made about the SCP
projects, having no technical expertise which would enable them to
second-judge their executives' decisions. The projects were created,
maintained and eventually closed according to the judgements of
individuals in the R & D department, who differed in their views of
the value of these SCP projects. There was therefore no remaining

SCP project by the time SugarCo went through its 1979–80 financial crisis, which was precipitated by a drop in the world sugar price. SugarCo came sufficiently close to bankruptcy that the non-executive directors called the banks in to examine the company's activities. Drastic action was necessary just to save the company. New Ventures was closed and the research department activity curtailed; employees were reduced from over 180 to 100. Research returned to a service function, with the exception of the experimental synthetic sweetener; this continued in development until sold off in 1992. The new head of fermentation commented that if he had not closed the SCP projects they would have had to go in this period of retrenchment.

Commercial scrutiny in the R & D department

The first head of fermentation referred to the change that occurred when Finlay had become a member of EXCO, as having the effect of at once increasing the resources available to R & D but bringing also an increased commercial scrutiny of projects. However, the next head of fermentation described SCP as a typical example of 'technology push' without hope of commercial development. When comparing the department during the period 1970–9 under Finlay with the department after its retrenching reorganisation, both commented that the old department appeared far less commercially orientated than the new.

The R & D department as run by Finlay published a series of booklets called *New Ventures*, which detailed the department's activities. The head of fermentation at that time pointed out that the function of the *New Ventures* booklets was to show that the board were getting value for money; that research was 'cost-effective'. In the absence of other criteria, it did have this effect

> in its broadest sense. They were spending £x a year into a kitty called R & D, the research organisation had to respond by showing that that money had been spent wisely. Because those who controlled expenditure were unable to judge technical matters the sorts of things that the R & D people did appeared to be cost-effective. That is why the *New Ventures* booklet thing arose, this was Finlay's response to them saying, 'We're giving you £2m/year, what are you doing with it?' . . . It served a real purpose of communication but a secondary one of being a very good PR job. . . . Cost effectiveness means people believe that the money they are spending is spent to good effect.

Although the board's approval for the large-scale funding of research was necessary, the structure and membership of the board allowed the R & D department to develop its own strategy for diversification. In hindsight this was insufficiently market orientated for the company's needs. In the nine years that the 'diversification through R & D' programme ran, no commercially successful projects emerged, and with the sale of Talin the last project generated from this period with a change of major sales has been lost to the company.

THE SHELL SCP PROJECT AND SENIOR MANAGEMENT

In the 1960s Shell was seeking new uses for the natural gas to which it had access in the North Sea. The availability of large supplies of gas was the principal reason why Shell chose to research an SCP process which used methane-consuming bacteria. At the time there did not appear to be alternative uses for North Sea gas and long-term cheap bulk supplies seemed assured. Hamer (Shell) recalled that the obvious use for natural gas was as a fuel to replace town gas, but there was no confidence in the management ability of the UK gas industry to change from town gas:

> Town gas went out through the 1940s and 1950s, they were losing their share of the market daily. So they were in a dreadful state. The gas industry was a disaster . . . the marketing of gas and the business was an absolute disaster. So companies like Shell really felt they had to find some other way of selling their gas. [There was] no confidence in gas.

In the 1960s natural gas was being flared in enormous quantities in the North Sea and there was even doubt whether the technology for laying long pipelines to distant offshore fields existed.

The other reason why Shell opted for natural gas over paraffins was its lack of secure access to paraffins and its possession of an industry based on paraffins for the production of biodegradable detergents. This industry had built its paraffin market knowledge over years and its internal forecasts predicted rising prices in the medium term. So Shell saw no future in paraffin-based SCP processes at a time when BP was becoming fully committed to them.

Research at Shell's Sittingbourne site had to be supported by one of the Shell operating divisions or by Corporate Research. The SCP project was first supported by the National Gas Division, then Corporate Research, then Chemicals Division. Each year the division responsible for funding would review the project and decide whether to continue its support. The Natural Gas Division was interested in SCP as a possible use for its large reserves of natural gas, but in the 1970s, when the market for natural gas as a fuel was developing, the Natural Gas Division lost interest. Corporate Research then took it over because of a strategic interest in biotechnology, then the Chemicals Division acquired the project because, according to a research manager at Shell,

> it was trying to diversify at the time, to look for new activities . . . the operating divisions act independently in that way. Chemicals, after a learning lag, costed the process, which neither Natural Gas or Corporate Research had bothered with.

This report found that SCP was unlikely to be economic unless a partner could be found to bear further development costs, and the project was closed in 1975–6. A project manager commented that of all the hydrocarbon SCP projects, 'In retrospect, Shell probably made the best decision, from a commercial point of view'. The relations between the R & D personnel who ran the project and the operating divisions' management who funded it were described by a research manager:

> Links between R & D and management here are very weak . . . they have been. We have moved to justifying R & D more, priority rating of projects so that in times of little capacity, we can save the most useful projects. The organisation of responsibility had a diluting effect on hard commercial analysis, and the costing [of SCP] was delayed.

One of the research scientists described this weak relationship between R & D and senior management in general terms:

> We have the *New Scientist* syndrome. . . . Fishlock[2] might write something in the *FT* and we get these articles sent down from on high . . . with comments scribbled in like, 'Why aren't you working on this area?' Some messages on PHB [a biodegradable plastic] came down recently . . . in fact, the easiest way to get approval is to have something published in the *New Scientist*. We have to sell our research, looking at it from year to year. They look for profit in the short term. In Research, we have a duty to provide options for the longer term. You must have faith and believe in it yourself, and yes, do what we believe in and sell it to the divisions.

The research department picked its own projects and researched areas of general interest to the research scientists, attempting to persuade senior levels of management that these were worth supporting. This meant senior management did not have an in-depth knowledge: 'Senior management don't know a thing about what's going on in Research . . . if we were all sacked tomorrow it would make no difference to the profits of Shell'.

A number of managers made the point that Shell's core business did not depend on successful research; this remained a peripheral activity.

> Shell doesn't make money out of research. It makes money out of holes in the ground, that's its business. And it's made money that way for so many years, that research, OK, pay a subscription, keep a few scientists off the streets. . . . ICI are a research-based company, that's the difference.

For Shell, research insured against change in their main business:

> It is preparing for change. What happens if the equation goes that in chemicals there is all the added value and all oil production is nationalised? What do you do? That is why they diversify into gas and metals.

One of the SCP project leaders thought that the practice of siting development in Amsterdam could have affected the SCP project, albeit in what he saw as a negative manner:

> basically, all development, almost by law of the company, has to be carried out at Amsterdam. So research that is carried out at Amsterdam sees development at Amsterdam, research anywhere else never sees the light of day, they will never put it in the development phase, because the Amsterdam labs believe all process research should be at Amsterdam. [SCP] was always going into development, but they always found excuses to slow it down, it was always more important to do something else. They never gave it high enough priority in the management.

The Amsterdam laboratories did not do much biological research,

> because they said it would be competing with Royal Dutch Fermentation industries, Giste-Brocades. Alright, Shell and Giste-Brocades have a number of tie-ups now. But, 'We cannot

compete against our friends in Giste-Brocades'. I've heard them say that, at senior management level. They used to have a microbiological department in Amsterdam, who did nothing. But again, because it would be competing with Giste-Brocades, it was closed down, about 1970–1.

A former Sittingbourne SCP researcher who subsequently moved to ICI commented that

> You have to recognise that the R & D facility down at Sittingbourne was really a drop in the ocean for Shell. It was just a tax benefit. Really, whether the thing [SCP] went on at Shell or not was neither here nor there, unless they were going to commit to build a big plant. As far as I know they were nowhere near making a decision to do that. I think its a sad reflection on the research at Shell that the research in biotechnology has been going on for a very, very long time, 20-odd years now to my knowledge – not one single product has come out of it.

This manager acknowledged that Sittingbourne was run in a different way:

> there was a hell of a lot of academic freedom in the Sittingbourne research labs. ICI were totally different and BP totally different too. There was a major thrust on projects and you were allowed very little academic freedom. That I think was right, and that is why Shell never got anywhere with anything in the biological sciences.

When SCP was closed the people who had been involved looked for other projects that they thought might be useful to Shell and that were concerned with their fermentation and microbiological skills.

> After 1975, there was a period of trying different projects, a good five years before we settled down. The next sort of buzz thing in biotechnology was the oxidation of propane, petrochemicals and biopolymers. Now, we have a better understanding of the role of biotechnology in Shell. We choose what we think will be of value to the company . . . no manager has the information to decide amongst the biotechnology projects . . . nearly all the biotechnology research is bottom–up research. We are pretty independent of company strategy.

The SCP project in Shell did not come near to being developed into the full-scale plant stage. However, researchers and managers at a lower level deduced an informal senior management strategy from various events and patterns of behaviour in the company, a strategy which was not favourable to taking biological research into the development stage. In the 1970s at least, the relationship between R & D and senior levels of Shell management was seen by some SCP researchers and managers to be minimal and the board as remote.

HOECHST AND A GOVERNMENT 'TECHNOLOGY STRATEGY'

The Hoechst SCP project began in the company's pharmaceutical division, where a senior research scientist, Prave, saw the rate at which BP was advancing and started his own work on yeast SCP. By 1971 Prave was looking for a company to help scale-up to the pilot plant stage and chose Uhde, a plant construction company owned by Hoechst. Financial support was obtained from a government agency, the Bundes Ministeren Forsch und Technologie (BMFT). The BMFT had already been approached by a state-owned mineral oil company called Gildenburg, who wanted support for an SCP project of their own, and so the BMFT suggested the two projects join together. This was agreed by Prave, and the BMFT began to fund the joint Gildenburg, Hoechst and Uhde project in 1974. One of the research scientists on the project explained why the BMFT decided to support the project:

> support was around 50 to 60 per cent of costs. Because they [BMFT] said it could be a nationally important project, we lack experience that English companies already have, maybe the Japanese have, but no German companies. [Also] at that time, because of the protein gap, only 20 per cent animal feed was produced in Germany – you should look at national security so that in case of a shortage, as in 1974, there would be the opportunity to produce German synthetic protein. Very important, this, that it was seen as of strategic importance.

The support of the construction company Uhde was decisive in convincing the Hoechst board to support the project as well. The project continued until 1977, when the BMFT announced that it would cut its funding and it was up to Hoechst to continue if it wished. The peak in protein prices in 1974 had passed and it had become apparent that there was going to be no international protein crisis. Hoechst would not increase funding to make up the government's contribution and so the work declined to the value of Hoechst's contribution.

The project management reacted to the closure of the BP plant in Sardinia by switching from paraffins to methanol as feedstock. The consensus in Hoechst, as elsewhere in the late 1970s, became that paraffins could no longer offer an economic route to SCP production but that methanol was the route forwards. Hoechst chose methanol despite a lack of control over the supply, but they were attempting to develop SCP as human food and so were not so sensitive to the supply or cost of the substrate. One ICI source suggested that Hoechst's choice of the human food market was a result of their lack of control over methanol supplies, reinforcing the idea that long-term control of a substrate was an important element of every company's decision to research SCP as a bulk animal feed.

In 1978–9 the project was moved out of pharmaceuticals into the general research division because it had become so involved with animal nutrition and was clearly not linked to pharmaceutical research. Gradually it came to be seen as a project that would never sell and the director of the general research division closed the project after consultation with the project managers, scientists, and also with the ICI board of directors in view of their Pruteen experience.

The project can be seen as one in which the government was the most important controlling partner. The BMFT agreed to funding provided Hoechst-Uhde and Gildenburg worked cooperatively. The BMFT paid the entire cost of the pilot plant and attached its own technical criteria to its construction to enable it to be useful in other research than purely SCP. The project was supported for national strategic reasons, and without the government it would not have taken place, despite the fact that Prave had the original idea. Jasanoff (1985) considers that the BMFT funding of the Hoechst SCP project was one of the largest public–private undertakings in German biotechnology in this period to the mid-1980s. Although Jasanoff also describes the programme as technically successful, she considers it to be of questionable value from the BMFT's standpoint of wishing to place Germany in the forefront of biotechnological competition.

RHM AND SENIOR MANAGEMENT DECISIONS

The mycoprotein project was conceived by Lord Rank and Arnold Spicer, the RHM director of research in the 1960s, to provide a protein-rich food to third world countries. The project was the brainchild of Lord Rank and after his death it passed through a period when Sir Peter Reynolds, Jack Edelman and A Spinks, as executive sponsors of the project, protected it during its early stages of development when it could easily have been closed for short-term reasons.

The project was run in collaboration with Du Pont for many years, but when Du Pont left the project in the 1970s the project managers had to rethink its future – if it was to be paid for by RHM it would have to face board scrutiny. At this point the project, under Du Pont's influence, had moved down the path of a low-cost, alternative source of protein for the third world, and the economics meant that board support was unlikely. One of the project managers saw Du Pont's withdrawal as positive in retrospect:

> In fact, funnily enough, it didn't feel like it at the time, Du Pont leaving was one of the more positive spurs to the project. It forced dramatic reappraisal of the project. It caused the science and commercial base to think again about where they were going. To see if there were any discontinuities in, for example, price–volume relationships, in the technology that might, might not be available.

The reappraisal did result in a change of direction in research strategy.

> It was about that time that fresh, frozen higher-added-value forms of the product, more suited to Western rather than global markets [were considered] and, if you like, the perceived value of what the project was worth dramatically improved. . . . Suddenly RHM had to think, solo, about what it was going to do next. The shock that we had no money . . . since that was a very strong part of the previous reasoning . . . if we continue, how do we make money? If we are going to make money we must sell it for rather more than we think we can sell it for at the moment. How do you do that? Well, you don't sell a powder, you sell a wet, moist form of the protein to Western Europe, not to the Third World, . . . suddenly there was just a sharpening up of the rationale in the period, 1972–4.

Once this new rationale had been adopted it was used by the project managers to successfully convince the board that the project was worth supporting – but they only approached the board once they believed it had a good chance of being supported. Advance knowledge of how the board made commercial decisions was essential in helping them to achieve a new vision of the project and the project that the board had placed before it was one that had already met the most important criteria by which it would be judged.

In the next few years the base technology for the new form of the product was developed, including such elements as a technology for retaining texture in the product. By the early 1980s there was a need for a pilot plant, but the board were not willing to spend the £4 million to £5 million required to build it, while the project had become more expensive than had been estimated in 1972–4 when the new marketing vision was adopted.

> we didn't have the money and we couldn't find a partner on terms that were suitable to us. . . . Had they understood the full financial significance of what they were doing, I think they [project management] would have found that decision [to support the project], much more difficult to make, much more difficult to carry their colleagues on the board.

Then in the early 1980s RHM found ICI with a spare pilot fermentation plant on their hands. According to one research scientist, 'rumour has it that their technical director met our technical director at a conference in Switzerland, or whatever, and they got together'. ICI and RHM formed Marlow Foods to coordinate the marketing of the food products containing mycoprotein. The research activity was split according to the perceived strengths of the two companies, process research taking place at ICI Biological Products, Billingham, and texturising research staying at High Wycombe.

Although the project was initiated at the highest level the board were not willing to fund development unless they could see how it would bring a commercial return to RHM. Until this time the project had to rely on outside funding and internal high-level protection. Ironically, it was the shock of the withdrawal of outside financial support that triggered the rethinking process, which in turn led to a rationale for the project that promised to bring a commercial return to RHM, and so to the board's approval for development funds.

LIQUICHIMICA MANAGEMENT AND GOVERNMENT INCENTIVES

Liquichimica was the subsidiary of the giant Italian company Liquigas. Liquichimica specialised in refining normal paraffins out of crude oil, fractionating the paraffins and selling them on to end users. The company was a subsidiary of the giant Italian company Liquigas, but the idea of using SCP to add value to the least useful heavy paraffin fraction arose within Liquichimica. Liquigas management were kept informed of Liquichimica's progress by Ursini, the president of

Liquichimica, but took no active part in the project. The SCP project manager described the company's strategy:

> They tried to be downstream integrated, in the sense of, normal paraffins, normal olefine, linear k-benzene and [then] petro-protein. The weak point was that they were not linked to the company supplying gas-oil. They had to buy gas-oil to purify and sell

Liquichimica was one of only two firms to use Japanese SCP technology. In Liquichimica's case they used Kanegafuchi SCP technology under license, establishing the agreement before the Japanese government banned licensing. The agreement was facilitated by Liquchimica's extensive Japanese trading.

> We were supplying paraffins mainly to the Japanese area at that time. Through Mitsui. So we were familiar with the big Japanese companies like Subito and so on. We also bought our synthetic fatty acid technology from Nippon-Soda. There was a particular intensity of relationship at that time.

Liquichimica had intended to build their SCP plant adjacent to their paraffin-refining plant in Messina, Sicily.

> Then the government asked Dr Ursini to invest in some southern Italian area. But to build the plant on the mainland required auxiliary services, steam production, energy, water and so on. This was not a feasible investment with just one plant. So we looked for other projects over which to distribute the fixed costs . . . and [we thought of] the citric acid and fatty acid projects – the citric acid plant was only built because we were determined to make an investment [with a return]. The port facilities, everything had to be built from scratch.

The heavy fraction of paraffins was considered a waste product by Liquichimica, and the original justification for SCP was based on a zero-cost raw material and an SCP plant conveniently sited next to the fractionation plant supplying its raw material. Now the integrated plant was to be built near Reggio di Calabria, 20km across the Straits of Messina and 20km from the paraffin plant. The construction of the integrated plant was heavily subsidised, but the project manager thought that Liquichimica was exposed to an investment of between 150 billion and 200 billion lira, by 1973.

> They paid us, that's why we did as they wanted. The citric acid and fatty acid plants were feasible with government funding . . .

Ursini was a financial man, born here. He accepted this role we were to play with the state, to affect the social conditions of the south of Italy.

The SCP project manager felt that for Ursini,

probably his mistake was to approach this high-risk project too rapidly. I think there is a co-responsibility between the government and the technician people – the technicians with a petro-chemical mentality, not an industrial microbiological mentality, gave Ursini dimensions of a plant that were too big.

Liquichimica had almost completed the plant when the 1973 oil price shock arrived.

So they decided to complete [construction] and start a very wide-ranging project of research on toxicology in order to be ready to demonstrate that this product was also safe for human food. I was supervisor of this project

The company strategy was to escape competition with soya, and this manager felt that it was the rumour of this intention to move into human food that sparked the anti-SCP consumer movement in Italy. The rumours were well founded, which is perhaps why Liquichimica refused to discuss them with the media, and why the media distrusted every toxicological claim that BP and Liquichimica made for their products.

Liquichimica also lost valuable time waiting for a serious mistake by the public health laboratories to be corrected. The Istituto Superiore di Sanita (ISS) found that the Liquichimica strain of yeast was pathogenic. Liquichimica thought that the ISS had mistakenly contaminated the Liquichimica strain with a pathogenic strain, Candida Albicans in the government laboratory. The ISS was eventually shown to have been mistaken by various independent experts who repeated the tests on Liquichimica's yeast strain. At one stage the ISS also claimed that Liquichimica were using the strain Candida Tropicalis, which was untrue. To clear up these confusions took two years and as the delay drove Liquichimica closer to bankruptcy, the president of the company decided to sue the ISS for the production lost between 1975–7.

Liquichimica attempted to deal with the problem of the SCP plant not being licensed to produce by rescheduling the construction of the three plants.

Liquichimica had planned to have the plants completed in 1975 for SCP, 1977 for citric acid, 1978 for fatty acids. When the approval date was seen as unforecastable, they changed the dates – to finish citric acid immediately and start manufacturing it. The problem was the utilities . . . [which] were engineered to cover three working plants. It was uneconomic to run just the citric acid plant with the fixed cost of the utilities. We ran it for one year producing some 20,000 tonnes of sodium citrate.

But the delay continued for three years and Liquichimica finally went bankrupt in 1978, and a commissioner was appointed by the creditor banks to decide whether the company could be saved in some form. By 1979 this commissioner had formed the opinion that approval would never be given to their protein product, Liquipron. The government then transferred the Liquichimica fractionation plant to ENI, which it operates as before. Now the integrated plant, complete with port facilities, remains derelict in Montebello di Calabria, near Reggio di Calabria.

The Italian health authorities' refusal to license SCP production was therefore the immediate cause of the write-off of this huge investment of largely government money.

ICI AGRICULTURAL DIVISION AND THE MAIN BOARD

A Pruteen manager described Agricultural Division's 'methanol culture' in the early 1970s:

If you were coming up with innovative processes then there was a culture that said if you couldn't make it from methanol then it was not something for ICI . . . and there is a very famous story about this . . . there was a discussion with Hart[3] with the researcher saying, well, we've been beavering away on a new process that we haven't told anyone about, we've got this microbiological process for turning lead into gold. And Hart's response to that was, well, that's all very interesting, but can't you do it with methanol? And that was the sort of culture you had.

The same manager made it clear that the 'methanol culture' was based on commercial reasoning, and the choice of methanol for SCP was because it was the most suitable substrate for SCP at the time. The importance of this 'culture' was that it was used to select viable methanol-based projects over other substrate-based projects, whether or not these were viable.

ICI had negotiated a huge contract for natural gas from British Gas in the early 1970s for a very low price per therm. Managers in other companies believed this privileged price contract was the reason why ICI chose methanol for SCP, but this view was not supported by those who had worked on the Pruteen project. One ICI manager denied that this was the case, because ICI had undertaken to pay for the conversion of the UK from town gas to natural gas as part of this contract. So the apparently advantageous contract was not so cheap as it appeared, although as a result ICI had a major incentive to develop new uses for natural gas and methanol.

Before the idea of Pruteen arrived some managers saw Agricultural Division being used as a cash cow by the main ICI board, with one commenting that

> For years they milked Agricultural Division and the result is a total bloody mess. I think the last plant to be built was over ten years ago. Things were allowed to go on as they were because of the money it made. £100m profit a year. They [Agricultural Division] got precious little back from the centre . . . nothing.

Others made similar comments: 'Yes, in those days fertilisers were churning out about a third of [total ICI] profits . . . the amount of money being generated . . . why not spend a little of it here?' So there was a background of perceived 'investment neglect' within the Division. The Division board was responsible for the design parameters of the Pruteen plant and the decision was commented on by a marketing manager:

> In those days Agricultural Division was very much a science and engineering company. Marketing did not play a very big part. The vast majority of the employees and the senior managers and the directors had an engineering or research background. Basically I think they got mesmerised a bit by the benefits of scale. For myself – in hindsight, I would have been right – I would have gone for a 25,000 t/a plant where you could have been confident that you could sell the output. The engineers convinced themselves, and they had the power, that to get the unit costs down, you needed in excess of 50,000 t/a.

Another manager described the divisional board as without sales people or even many scientists. Those that were there 'bitterly resented the control of the chemical engineers. There were not enough voices to create a debate'. In 1974–5 when the design parameters of the Pruteen plant were set, 'a decision could have been taken to stop the development of a 50,000 t/a plant. Instead the debate was over building a 100,000 t/a or a 50,000 t/a plant'. A

100,000 t/a plant was obviously more expensive than the smaller plant envisaged at this time. But as one manager recalled,

> Even with the extra capital costs there were a whole lot of arguments around; many items are standard, it isn't that much more, we were in a development area and there was 52 per cent free depreciation and 20 per cent development area grant. The amount of cash it cost you was only something like 28 per cent of the cost. All these arguments went along with a big plant. I went along with it at the time. I didn't have the power to do otherwise, but in hindsight it should have been a smaller plant. The argument against the smaller plant was that if you are going to build a world-scale business in a market of tens of thousands of millions of tonnes of Pruteen, you don't build tiddlers; if ICI were going to be in, it should be done properly.

The argument for the smaller plant was that this was of a capacity, 25,000 t/a, which suited the size of the veal calf milk replacer market. In this market Pruteen would effectively replace skimmed milk powder, with which Pruteen was, at first, competitive (see Chapter 3). However, skimmed milk powder even then was being subsidised by the EC for use as animal feed. In Agricultural Division most managers believed that this was a temporary phenomenon. According to the toxicology manager,

> At the time our colleagues were writing the capital expenditure proposals for the Pruteen plant the skimmed milk mountains were only just starting to evolve. These are now permanent features of the economy of Europe. We assumed that the mountain was a temporary feature, it would disappear and we would be able to obtain a premium over and above fish-meal because of the price of skimmed milk powder.

He believed that it was ICI's lack of experience of how the EC behaves that allowed the board to keep this belief that skimmed milk prices would be well above those for Pruteen. This is supported by the comment of one of the Hoechst project managers, who recalled debating the issue with ICI:

> And we said, there is no way of seeing the [EC] give up influencing the skimmed milk price. A difference of opinion! We never thought to compete with skimmed milk powder. With the agropolitics in the EEC.

As discussed in Chapter 3, other managers in ICI and RHM disagreed with the Agricultural Division assessment of the soya market at the time of the ICI main board's decision to approve the Pruteen plant construction. One of these gave a more caustic view of how the Agricultural Division board was motivated when they proposed expenditure on the Pruteen project:

> They wanted reasons to develop this technology because they believed in it. They believed it was the basis for the future. And if it hadn't been Pruteen it would have been . . . they would only have had to find something else. So I think there was a pre-conditioning that said, look, the last thing we want to find out is that this is not a very good thing to be doing at the minute. Because it makes life a bit tedious when we think about the technology we will need in the twenty-first century. Just find the reason that says the thing is sensible and get on with it.

Elements of this idea that the Pruteen technology was worth supporting for its own sake was supported by at least one Pruteen project manager, who described how he and others in Agricultural Division saw the technology:

> they'd recognised fairly early on – I think this is very important – that ICI's process was based around a very novel piece of technology, around a novel fermenter design which was called the pressure cycle fermenter, which ICI patented. They'd had that design for a number of years in the early 1970s. It was an extremely important piece of technology . . . what ICI recognised was that in the longer term biotechnology and fermentation processes were going to be needed on a very large scale. So there were two processes driving it really. One was the technology, the fermentation technology they had developed, and the other was the process itself.

The capacity and other design features chosen by the Agricultural Division board was turned into a detailed design document and submitted to the main ICI board for approval. Agricultural Division worked with John Brown Engineering to prepare the plant design. One of the John Brown managers who worked on preparing the division's proposal described how it was eventually approved by the main board:

> it took some time before the board said yes. ICI came here [John Brown, Portsmouth] and we set up 200 people in a team to do this very complicated design engineering, and six months later the board said no. That was in 1974. So the whole team was disbanded, so every one went home. Six months later ICI said, well look, send

up just six of your people, who were process engineers and have another go, devise a revamped design. It was during that time that the size of the plant came down and certain other things with regard to efficiencies came down . . . from 100,000 t/a to 65,000 t/a, which is what that plant is capable of doing . . . we made it cheaper in one way or another. And they said no again . . . well, a year after that, two years after we first started, the board gave the go ahead. [That was] 1976 . . . three years later we commissioned the plant. Now that was pretty fast track for new technology and such a large plant.

The reason for the delays was explained by the toxicology manager:

You are dealing with a decision by the main board at Millbank, making comparisons between projects from other divisions. It is whoever gives the most convincing stories at the time, combined with the strategic view of the Chairman that decides what happens. A project may be told, come back in 12 months time . . . if you still think you want it, then maybe. There are a whole lot of reasons why the actual decision taken can vary on a time scale which could be completely independent of what an outside assessor might think is the actual investment case. You need to go back and look at the company reports, starting in 1971–2, look at the capital investment projects the main board has supported and also the profitability, which comes from different sectors of the company. Put the two together and then make your own judgement.

Another manager told how the main board directors informed themselves of the background to the proposed Pruteen plant:

Virtually all the executive main board directors came up on their own and spent a day [at Agricultural Division], during which they received a presentation from us. They had opportunities to probe as much as they liked . . . there was plenty of opportunity at that stage for the directors to get to know what they were letting themselves in for.

He told how the main ICI board functioned in the 1970s and up until Harvey-Jones's reorganisation in the early 1980s:

It was still that main board directors had responsibility for one or two divisions, or areas of business. And Swarthers[4] [member of the main board] was really our link between the division and the main board. A lot of the discussion was really with him. If he approved it, then . . . in fact my directors, Hart, then [later]

Trowgers, were able to bypass the board [of Agricultural Division] and go straight to Swarthers.

He went on to describe one of the results of this system, superficially similar to SugarCo's:

> You see, people like him [Swarthers] tended to be product champions. Once the likes of him thought, we'll run it for another year, and we'll try and achieve this, once he was convinced that that was what we were going to do, he could peddle that . . . there was no dissent [on the main board]. . . . Now the way ICI's organised, its much more dependent on a business manager reporting to a small sub-group of the main board, with no one too committed to it. . . . I think it is healthier now, [projects are] reviewed with some frequency, so if they are not doing as they should, they can be changed.

Swarthers continued to be the main link between the main board and the Pruteen project in the first few years after the completion of the plant.

> During the first two to three years when we believed we could build big plants or license . . . Swarthers was supportive, but I guess he didn't expose too much what we were doing to the board. It would have got maybe a couple of half hours a year, kind of thing.

During the first years after completion of the Pruteen plant in 1979, Agricultural Division managers appeared to remain optimistic about the project's future. One manager commented that 'In 1981–2 they were planning "Pruteen 2" . . . 250,000 t/a!' The Agricultural Division board were still looking to future economies of scale because it was recognised that the plant that had been built could be improved upon second time around.

> Technically I think it was possible to go from 1.9 tonnes of methanol per tonne of SCP to 1.3 tonnes. Technically, it was possible to use less refined methanol. There was not just the cost of methanol but of drying and so on. The benefits of scale . . . we certainly went through a phase when we believed that with 300,000 t/a plant we could produce good businesses provided we were careful about the market we went to and where we sited them . . . but all that was still based on high soya.

These mega-plants were never built: 'The decision not to build a second large reactor was one that took itself; it became self-evident that it would not be viable'. At the same time as faith in the future of Pruteen was waning, the incentive to find alternative outlets to the animal feed market was growing. When asked at what point human food research began, the toxicology manager replied,

> At no particular time. It just became obvious, yes, a gradual evolution . . . at no time did we say, right stop researching animal nutritional properties of Pruteen, start human food research.

Agricultural Division took a number of decisions 'against the grain' when they decided that Pruteen had a future in the 1970s. With a background of grievance at not having had fresh investment for a long time and seeing in Pruteen a desirable base technology for the future using methanol, the feedstock of the future, they interpreted changes in key price factors in a manner that favoured advancing the project; they believed the soya price would stay high, apparently against the expectations of both the feed industry and ICI crop specialists (see Chapter 3). They believed that the EC would not continue to subsidise skimmed milk prices, when Hoechst were arguing that the EC would continue. The economics made sense for a time, but the future depended on the interpretation of these trends in prices of inputs.

The project was approved by the board, but from the way the board operated at that time it appears unlikely that it did more than to take Agricultural Division's projections on trust. That is not to say that Agricultural Division misrepresented figures to the main board – they undoubtedly believed the project to have been worthwhile – but the main board could have been organised in such a way as to enable more rigorous scrutiny of the Agricultural Division thinking, as is now the case.

The second oil price rise broke the economic case and between 1980 and 1985 the emphasis on the project shifted to development of the fractionation project and an effort to license the Pruteen production technology to other countries.

Reorganisation and the end of Pruteen

The realisation that the Pruteen project was not going to expand rapidly and that the animal feed market was not going to materialise appears to have come over a period of two to three years; there was no single point at which Pruteen came to be seen as a commercial failure. In this period between 1981 and 1983, a manager commented on the main board's view of Pruteen, 'I don't think they knew what to do with it'. Finally a new manager of the Pruteen business area was brought in,

> Trowgers was brought in as general manager of Pruteen, but then within about a year and a half he was made a director, but he also got [Agricultural Division] research. He had two groups of people, Pruteen and research. Research were more or less divided into catalysts, support of ammonia, methanol and biological.

Trowgers reviewed the future of Pruteen along with the research responsibilities, as one of his managers comments:

> We started from Jon [Trowgers] being an optimist. When we realised we were not going to make it in the animal feed market, his reaction was, Christ, we can't throw away all these skills. The least we can do is to see if anything can be done with the plant. . . . Although it didn't succeed, we still had a lot of knowledge of how to develop and evaluate products and to get in and understand markets. We should not throw that away.

He described how Trowgers saw the existing structure in Agricultural Division:

> we had big research, big engineering departments, quite quickly we realised they were working on the wrong things. When I got involved there were 31 projects. On some of them there was little going on. But you could sit down and list 31 projects in what was the biological part of the research department. What we needed was a structure that would get that much more market orientated. That would begin to reduce that 31 down into six or eight strategic areas, and get some criteria around which you could judge those.

The process of thinking out what these criteria would be was confined to a small, informal group.

> Basically he [Trowgers] was using the few other people he trusted to help think how we could make this thing more effective. What was apparent was that in this day and age you could not have a free-standing research activity that set its own targets . . . the commercial input they had was a couple of blokes from planning and coordination, who looked at the market and assessed the market . . . but there was no real executive authority – that resided with the research manager and they had to fight for his money each year. But a lot was being spent on things which never had a hope in hell of seeing the light of day.

Asked why, if the planning people were responsible for assessing the market, there was nevertheless expenditure on such non-commercial projects, he replied that

> they didn't [define the market]. It was a cosy club. They were all my good friends . . . and still good friends . . . who sat around review meetings once every two months and reviewed all the group projects. But there was not enough steel or hardness about it . . . there was no one saying, where is this taking us? If we spend £200,000 in the next year, where are we likely to be and what are the options? That may seem odd for a company like ICI, but it is typical of British industry. I think we did a fair part to sharpen that whole thing up.

Trowgers had the problem of how to implement his ideas for making the Pruteen business area and the biological area of research more market orientated.

> Now there was great resentment against actually imposing this kind of thing. But we won them round. What you've got to do is for people themselves to see that this isn't actually going to make anything. If you can get some [agreed] criteria for whether a project has any chance of succeeding, you get people to close down their own projects. But in a way that does not demoralise them, so that they will still look for new things.

An event was arranged to help persuade research managers of the need to change:

> we arranged a five-day session with Manchester Business School around the Pruteen project . . . on the process of innovation. And we got through to a lot of managers on my level or a bit below, what the mechanism of innovation is about. You have to have a hell of a lot of ideas, you have to allow people free thinking, but then you had to get rid of most of [the ideas]. And you had to spend a little money on a lot of things. You had to keep doing tests, technical or marketing, without spending too much money, to actually see if this is something that could grow [or] that fits with your strategy. And you've got to look at maybe 100 to get two or three that are worth putting £250,000 into. And you shouldn't be too hard at that early stage. But people have got to realise that when you spend real money you must be hard. With £200,000 and eight to ten people working full time, you have to believe something worthwhile is going to come out of it at the end

. . . at least the macro sums must be looking OK. When you move to spending between £2 million and £3 million, you must be even harder. People . . . hate their pet projects being dropped. But you've got to persuade them that they should drop them.

The Pruteen Business Area, which had been formed in 1974, finally became part of the new business area, Biological Products, which was created in 1985. Until the wind-up of the Pruteen Business Area the Pruteen plant ran in 'campaigns' where it produced Pruteen at well below its design capacity for the calf veal milk replacer market alone. This probably covered operating costs but certainly did not give a return on the capital invested.

In Chapter 3 one manager was quoted as having the belief that the 1986 fall in the oil price may once again have made Pruteen economic. The Pruteen toxicology manager independently recalled that in about 1986, when he left ICI,

> there was a concept last year or the year before that the problems associated with the economic production of our plant were being removed because of the fall in the price of hydrocarbon fuels. The strategic position remained the same however. Yes, it would be economic to operate the Pruteen plant in 1985, 1986, 1987, but ultimately the prices were going to go back to what they were. So we would spend a lot of management time wooing back the customers that we had originally had, they start to buy our products and then in five years time we walk away from them. It's not exactly the way to win friends.

The oil price fall came a year after the rationalisation of the Pruteen Business Area, a year too late. This time around ICI believed the drop in hydrocarbon prices to be a short-run hiccup in a long-term trend of rising prices; the interpretation of trend was all important, as it was in the decision to build the plant. The real legacy of Pruteen is in the technical expertise that was retained in Biological Products and the joint project on mycoprotein with RHM.

A similar conclusion can be drawn on the role of marketing input as in the SugarCo case. A planning function with market responsibilities existed in Agricultural Division research to advise on the commercial nature of research projects, yet the aim of the post-Pruteen restructuring carried out by Trowgers was to introduce stricter commercial control into research. It would seem that, despite the existence of a discipline provided by thorough project selection and evaluation procedures, there is no absolute standard for judging when research is sufficiently market orientated. That judgement is strongly contingent on the circumstances of the firm and of individual projects. I would suggest the general lesson is to avoid taking company claims that they pay sufficient attention to market input at face value but neither to be tempted to make glib conclusions about reasons for failure involving 'insufficient attention to market needs', at least without a serious attempt to investigate the criteria by which the firm organises the input of market understanding into research projects.

THE BP BOARD AND TOPRINA

In the 1950s and 1960s BP had large interests in Libyan oil-fields, which yielded a crude oil with a very high n-alkane content. The first fractionation of crude yields gas-oil, or diesel fuel, but if the paraffins are not removed before or after fractionation, they remain in the gas-oil and freeze at low temperatures, which threatens to prevent diesel engines working in winter. The initial attraction of SCP for BP was that they found yeast preferentially consumed n-alkanes in the Libyan crude, offering to solve the n-alkane extraction problem at the same time as creating a useful byproduct. Only later did production of yeast become an end in itself.

Unlike Shell, BP had no developed industry based on n-alkanes so they were actively looking for new outlets for these chemicals – they did not want to have to sell them to Shell, who did have uses for them. Shell also owned the world operating rights to a urea dewaxing process which was their method of extracting n-alkanes from crude oil. BP was reluctant to license this technology from their rival Shell and so there was another incentive to use yeast to 'extract' n-alkanes from crude oil.

Most research and development in BP is done in major stand-alone sites, such as the Sunbury Research Centre, and the BP businesses award contracts to these research centres to solve particular problems. There is also research funded by BP International, which is more strategy related and directly under the control of the board; the funding of research occurs within a structure similar to that of Royal Dutch Shell, but in the case of BP it was the corporate body, BP International, that funded Toprina from the early stages of the project in the 1960s until the creation of BP Proteins as a separate BP business. The BP main board actively supported and pushed Toprina along the path to commercialisation, in contrast to the Shell and ICI main boards. This is underlined by the absence of a Toprina 'champion' at a level in the company lower than the board. The most senior manager of the Toprina project began as project leader at Lavera, BP France, and became general manager of BP Proteins in 1978. He was never referred to as other than a competent manager doing his job and survived the demise of BP Proteins to become general manager of BP Chemicals in France.

Outsiders working in SCP commented on why the board should have identified with this development so strongly. For example,

> It [Toprina] was a flagship for their getting into something other than oil. The top management identified with it. No question, it had identity at the top of the company. Marvellous publicity for them. This was all before their big discoveries on the north slope of Alaska.

The company had decided that it wanted to diversify away from oil and SCP was the first step in what was to be a long-term strategy of developing associated businesses such as BP Chemicals. In the early days BP saw itself making money from a project that would help feed the protein-hungry world and it only made sense to milk as much positive publicity from this venture as possible.

The first BP SCP plant was based in their refinery at Lavera, near Martigues, southern France. The researcher in charge of the project, Champagnat, developed two routes for producing SCP, one based on alkanes and one on gas-oil, because he believed that SCP would be so important that two routes would eventually be needed and used. The board accepted this thinking and eventually approved the construction of both the Lavera plant, based on the gas-oil process and the Grangemouth pilot plant and the Sarroch plant based on the n-alkane process. The development of two processes and two pilot plants again demonstrates the company's commitment to this technology from early in the 1960s.

According to a former BP consultant who worked with Champagnat, BP barely paid any of the capital cost of the Lavera plant, these costs being borne two-thirds by the EC and one-third by the French Ministry of Agriculture. It was this gas-oil technology that BP succeeded in licensing to the Japanese company Kiro Hako. However, the plant suffered from bottlenecks in production throughout its life, and the former BP consultant at this time thought these technical problems persuaded the main board to make Lavera a semi-autonomous profit centre in the early 1970s. The plant was unable even to cover its costs with the 1973 oil price rise and was closed in 1976. The more important process was the paraffin-based SCP technology, which was developed into a full-scale plant sited in Sarroch, Sardinia, Italy.

All BP interviewees were surprised by the BP board's choice of Italy as a site for the Toprina plant. One remarked that the choice was

> Amazing. The funny thing is that BP Oil had just withdrawn from the Italian petroleum market, when BP International went into this joint venture with ANIC [the Italian chemicals combine]. The reasons [for the Toprina investment] were that BP didn't feel able to [pay for] a full-scale plant, because they were committed to megabucks investment in the North Sea, they were just about to open up Forties. There were plans to have one built just down the road, at Grangemouth. In retrospect, a much better idea. But they decided to go to Italy for the joint venture and extremely attractive development grants.

Another manager outlined some of the attractions of Italy and Sarroch for BP.

> the biggest single Italian refinery in Italy, Sarroch refinery, was owned by ANIC, which was part of the ENI group. Now it's ENI Chemicals. The feedstock to the Sarroch refinery was Libyan crude, very, very rich in paraffins. Sardinia was closer to Africa than to Italy, with deep-water port facilities. So from a geography and existing trade point of view, Sarroch was a very good source of paraffins. Next door to Sarroch refinery was Sarroch Chimica. It was the obvious place to refine the crude paraffins [from the crude oil in the refinery], take them over the fence to Sarroch Chimica, build a paraffins refining plant at Chimica to give you pure paraffins, then over the fence to Italproteine [the name of the jointly-owned, BP and ANIC, SCP producer company].

So both logistical reasons and finance were strongly in favour of building the two plants at Sarroch. In addition it was pointed out that

> Italy was a big market for feed. It was in the Common Market, which Spain was not. Things could have been different if Spain had been . . . but once a product is made in one country of the Common Market, it can go to all the others.

In those days before the oil crisis, BP made large returns on capital and large profits

> and again, we had the money to build it. If it had been now, it wouldn't have been done . . . once you start something and you have the money to go on, why not?

No one gave exact figures for the extent of the Italian state subsidies to the n-alkane refinery and the SCP fermentation plants, but one manager at the time stressed that they were large indeed, greater than 50 per cent of the total capital cost.

> On paper it looked good. On paper. The Italian government was able to pay for the project, because it was in the Mezzogiorno.[5] [They were going to pay] nearly all, it wasn't going to cost BP a lot.

Like ICI, the BP board were interested in SCP technology because they saw in it a long-term profitable future and many generations of plant. Had all gone well, future stages of development were envisaged:

> As one stage was starting to mature, the next stage was becoming obvious, being thought about and worked on. One used to say, in five years' time what will we be doing? . . . It would have depended on jumbo methanol plants in the Middle East for us. Then you would have had to build your plant out in Kuwait.

As with Shell and IFP, BP only gradually came to interpret the 1973 oil price rise as a permanent feature of the world economy rather than a hiccup of two or three years' duration. By this time, around 1975–6, most of the capital had been sunk in the Sarroch plant and the overriding reason for wanting it to go into production was to increase the possibility of it being licensed to other countries. A manager party to the economic analysis post-1973 thought the economics were not hopeless; it all depended on which economic scenario was adopted, and there were a number of these available to the board.

> The sums worked out to be a negative–positive balancing act, with the high feedstock price. I don't think anyone would have gone ahead with a highly negative product price. But there were these potential carrots for the future [the prospect of licensing].

At one stage BP had advanced plans to build a plant in Venezuela, but a condition for its purchase was the demonstration that the Sarroch plant could work. The interest being shown in the technology licence encouraged BP to hope that the rise in the soya price was not simply a reaction to the 1973 oil price rise.

> There was considerable interest being shown by the USSR and the Middle East, definite possibilities for licensing the process in both areas. I feel that is one of the reasons for their going ahead. I think there was a lingering hope that the soya price would keep on climbing. Wishful thinking.

Another reaction to the oil price rise had to work itself out before the situation crystallised as hopeless. This was the fear that oil supplies might be limited, in which case the company should diversify further. One of the microbiological researchers commented that

> At that time [c1975] they were wondering where BP was going. With these forecasts of oil running out in 1999. . . . There was a general worry that it might. It was all over the company. The diversification phase was the most interesting time for ideas. There were all sorts of things we could do with Toprina; we went through polyhydroxybutyrates [biodegradable plastic]; you could make practically any organic chemical. Or take the protein, dry it and make buttons from it. We looked at the semi-moist animal food market, to give to people like Spillers who would put it into Kennomeat meaty chunks. That folded up . . . we had a whole meaty chunk set up at Grangemouth that must have cost a whole lot of money . . . We were looking for other product areas . . . a very central idea, to diversify.

When asked what came out of the diversification effort, the reply was,

> I don't know, a good question. I suspect that someone did some sums, looked at the market . . . and knocked them on the head. Within the research department there has always been a group for economic appraisal. Either something can be knocked on the head by a simple bit of delving, or if its promoted a little bit more it will be knocked on the head after a bit of research.

Nothing did come out of this diversification exercise, but BP finished building the Sarroch plant only to run into a political quagmire with the Italian health authorities and hostile public opinion.

Clash of cultures in Italy

After the Japanese Ministry of Health and Welfare gave Kanegafuchi and Dainippon the 'administrative advice' to close down their SCP projects, the companies duly did so, in 1973. The reason for the government's 'advice' was a desire to mollify hostile public opinion, represented by two consumer groups who alleged that the Japanese SCP products could be contaminated with carcinogenic material. The Minimata mercury disaster and other industry abuses were responsible for creating this hostile environment, although the Japanese SCP products were never shown to contain carcinogens and were completely cleared by the Japanese government in the 1980s (see Appendix). Events in Italy now took a similar path to those in Japan.

BP completed its Sarroch plant in 1976 and Liquichimica was ready to begin production in 1975, but neither plant was ever licensed to produce SCP by the Italian Ministry of Health. The companies kept their plants staffed and ready to produce until 1978, when Liquichimica went bankrupt and BP gave up waiting for the Ministry's production licence and dismantled their plant. The three years that BP and Liquichimica were waiting to begin SCP production were a regulatory nightmare for the companies.

Many anecdotes were told of how difficult it was to do business in Italy, by British standards. So a manager from Shell remembered that when Shell was evaluating sites for its methane SCP pilot plant it considered the Superior Institute of Health in Rome:

> and yes, if we had been prepared to set up trust funds for most peoples' grandmother's aunts sort of thing, it would have worked. We actually became so nauseated, the backhanders that were wanted to make use of anything, that we said, no way.

There was little appreciation of Italian scientists either, one manager commenting that they frequently dabbled in politics:

> Most Italian scientists live above their means and are dependent on consultancy which comes in various forms, like making public statements about issues. There has never been any difficulty in buying Italian scientists . . . a different basis of ethics, because they are badly paid.

One former BP manager built on this widespread conception to assert that this was just what the opponents of the BP and Liquichimica plants in Italy had done. They had paid the biggest name in Italian microbiology to side against SCP and raise toxicological objections. These allegations could not be proved, but in Italy it was more than likely given the nature of Italian scientists at the time: 'There are some honest scientists in Italy. Some. I can say that, I'm half Italian'. On the other hand a number of managers referred to BP's ability to antagonise the Italian authorities. One spoke of BP coming to Italy with 'an oil-refining culture'. BP were 'used to dominating, "if you don't like us you won't get your oil from us", type of big brother attitude'. Another commented that

> the way the English behaved with the Italian administration was a disaster, a complete disaster. They considered the Italians as

> underdeveloped. They didn't consider small things like translating documents into Italian, because Italian was not an important language . . . it [SCP], worked in France, because the English were there as tourists, that was perfect, but in Italy the English wanted to behave like English.

The BP managers could not speak Italian either and in face-to-face meetings in BP Proteins in London they needed a translator, who was aware that this put them at a disadvantage. The Italians discussed the English in front of their faces and 'I was embarrassed, sometimes, to translate what they really said'. Where this clash of cultures became damaging to the project was in the relationship between BP and the Italian research centres, responsible for giving safety approval to Toprina.

A former BP consultant told the story of how a high-ranking official in the Italian veterinary service offered BP the permission to produce SCP in Italy in exchange for a new nutrition and toxicological testing laboratory, a suggestion which he saw as a demand for a bribe and which BP refused. One of the BP France managers of the time who had access to the negotiations confirmed that this offer was made, but in contrast to the view that this was a demand for a bribe and was outrageous, he felt that it was reasonable to help these research centres improve the quality of their equipment in this way. The head of what would approximate in English to the veterinary service was a very powerful man with the ability to halt the Sardinian plant, a power which he exercised. BP offended the research centre by criticising the quality of their analysis and persevered in ignoring the head of the veterinary service.

> The [BP] decisions were taken mostly in London and Milan. The head of the veterinary service was in Rome. Once he asked someone to come and talk, and they said, very sorry, we have no time.

According to this BP France manager, BP continued to believe for too long that the civil servants in the Italian Ministries had power, as they would have in Britain; BP did not understand how the Italian bureaucracy operated, and even tried to use the UK ambassador to pressure the Italian health ministries into approving Toprina. This further antagonised the Italians and this manager believed BP ignored the advice of their Italian partners, ANIC, who were aware of the problems.

The soya lobby in Italy

The reaction against BP and SCP in Italy was so violent that a number of BP managers thought there was some kind of organisation at work funding the agitation which they referred to as the 'soya lobby'; BP formally referred to the soya lobby as a reason for withdrawing from Italy in 1978. BP never publicly blamed the Italian soya importers, but some managers did think the soya importers were funding the public pressure groups and making submissions to the Ministry of Health that SCP was a danger to health. Where managers believed that Italian scientists could be bought and that a soya lobby was buying them, more extreme beliefs were possible:

> Soya was extensively grown in northern Italy . . . centred on Bologna. Bologna happens to be communist controlled. There was no way you were going to allow a fascist-imperialist joint venture in Sardinia put soya farmers out of work in communist Bologna . . . that lobby, the soya bean lobby, which was very strong, very powerful, very clever, got hold of this toxicological link, paid one of the biggest microbiological voices in Italy at the time to back their side, because BP were too slow in getting him to back their side. He backed them and shouted loudest.

One of Liquichimica's project managers objected to the idea of a powerful organised soya importer's lobby. He presumed the soya importers secretly funded the anti-SCP consumer groups, who were organising meetings up and down the country, paying scientific speakers and leafleting, all with no declared source of income. He felt that speculations about the existence and degree of activity of a secret lobby were only attempts to avoid having to explain the real hostility of Italian public opinion towards the SCP plants.

> It was a problem of public opinion . . . it was quite easy to say it was the soya lobby. The problem was that public opinion was not prepared to accept this kind of product in Europe . . . when it appeared to come from an unnatural source.

Another BP manager referred to the idea that a soya lobby interfered with the Ministry of Health as

> total speculation, just as it's total speculation that the mafia stopped the Liquichimica plant. No one is ever going to prove it. But when so many factions appear to be anti the BP plant, you decide, it just wasn't going to be allowed to work.

A BP France manager hostile to the conspiracy overtones of the soya lobby explanation, also believed that it was public opinion that stopped BP and Liquichimica, but that this hostility needed to be explained. A key influence was the banning of SCP in Japan in 1973. It was easy to understand that it had been banned in Japan, but it was difficult to understand the complex events which had led to this outcome, hinging as they did on the validity of toxicological trials and the trustworthiness of the legislators and companies (see Appendix). Because the reasons for the events in Japan were not understood, a much wider range of 'interpretations' of SCP were possible in Italy, including the idea that this disreputable technology was being dumped on Italy through the corrupt connivance of Italian politicians with multinational companies.

As discussed above, BP's original advertising of their SCP activities was well known in Italy and the press referred to Toprina as the petroleum beefsteak product (Il Sole 1975). All managers saw this as a colossal error in hindsight, one whose effect was to weaken the credibility of BP and Liquichimica denials that they were intending to market their products as human food products. In fact, a Liquichimica plant manager admitted that Liquichimica had been working on such products, but attempted to keep the work secret through fear of public reaction.

Both BP and Liquichimica adopted a secretive policy in reaction to the hostile public opinion, and this probably worsened the situation. BP's policy was to talk only to the necessary officials in the Ministry of Health and to work to persuade them of its case for the safety of Toprina. One manager who was strongly in favour of a more open policy at the time told how she was not allowed to make comments to journalists who managed to locate her and wanted to know the official BP line on Toprina's toxicity. There were no official company press releases on Toprina or the plant in Sarroch. In the absence of such information releases, the public debate on SCP remained lopsided and heavily against SCP.

The Ministry of Health

The Italian Industry Ministry could only issue BP and Liquichimica licences to produce SCP once the Ministry of Health had declared the two SCP products to be safe. A special committee of the Ministry of Health called the Consiglio Superiore di Sanita, or CSS, was given the task of commissioning toxicity research on samples of Toprina and Liquipron and reporting back to the Ministry. One of

Liquichimica's managers believed the decision to approve the products effectively rested with this committee and that the Health Ministry would accept its recommendations.

This institute never found the SCP products to contain any harmful residues, but did not make a recommendation to the Ministry that SCP was a safe product. Instead it repeatedly returned to BP and Liquichimica over a two to three-year period to ask for further tests and changes to their plant. Throughout this period the plants were complete and only awaiting permission to start production and managers in both companies believed the delay was deliberate,

> Because the decision would have been unpopular. Never any evidence of toxicity. Nothing. But they kept asking for more tests. . . . You see the [current] problem with toxic waste [Karina B]. I am sure public opinion pushes the politicians to treat the problem unobjectively. The type of experiment requested was the sort to take time, to allow a decrease of the anxiety of the people.

One of the Liquichimica managers had talked with many officials in the health institute and did not believe that any one person was really against the products. The decision to delay was political, and he implied that the Health Ministry might have influenced the outcome of the CSS deliberations as follows: 'if the Ministry say, give us conclusions to allow us to avoid giving approval. . . . CSS is managed by one [Ministry] representative and he is continually in touch with the Ministry'.[6] Liquichimica's view was that perhaps 35 members of the CSS were in favour of SCP and only three or four against it. This was enough opposition, because without a unanimous recommendation the Health Ministry was afraid to approve SCP for fear that the three or four members who were against it would go public with their opposition. A BP manager gave a similar account to Liquichimica:

> They produced objections, we countered them with fact, they produced more, we went to MIT, they had rent-a-mob, the local mayor at Sarroch, wanting to be re-elected, 'I will save you, the local population'. I think the last one, they asked for totally unreasonable effluent levels, by any industrial criteria. We kept on building, all sorts of things, but they kept on coming back.

These problems over the toxicology of the product did not occur in other countries. BP had been selling SCP from its French plant in Lavera for three to four years and ICI obtained UK government approval for Pruteen. No other European governments or testing laboratory found SCP to be unsafe.

The scientific debate over SCP toxicity in Italy

If the background to the refusal to approve the safety of the products was political, the language of the debate between BP and Liquichimica on the one hand and the institutes of health on the other was scientific.

BP and Liquichimica thought the alleged possible toxicity of SCP compounds made by some Italian microbiologists was based on disingenuous scientific reasoning. The result of the split amongst the scientific 'experts' was that the public and the non-expert officials in the Ministry of Health did not know who to believe and so avoided making any decisions based on one side or the other.

The first problem with the BP and Liquichimica products was that the n-alkane feedstock was derived from crude oil which was known to contain carcinogenic aromatic hydrocarbon compounds. Some of these carcinogens did survive the process that extracted n-alkanes from crude, but only in levels below those permitted in ordinary foods in Italy. It was alleged that when pigs ate SCP produced from the fermentation of alkanes, the aromatic compounds might accumulate in the body fat of the pigs, because it had been shown that traces of n-alkanes did accumulate in pig back fat.

In response to this argument the companies could show the results of toxicity tests which they had designed to show that aromatic hydrocarbons were not accumulating in quantity and were not a problem. The debate then moved on to the trustworthiness of the companies results and testing procedures.

Neither company had anticipated toxicological objections based on the safety of people eating yeast cell fatty acids. An ICI toxicological research manager thought this problem had proved to have been one of the most important problems for BP: 'This was used as a whipping post for BP. BP were stopped by arguments over the odd fatty acid chains. Personally, I have no problems over this metabolism'. The fatty acids produced by yeast metabolism have an odd number of carbon atoms in their carbon atom chains and these compounds were shown to be present in the meat of animals raised on diets containing SCP. People who ate the meat of these animals would therefore have to metabolise odd carbon-chain-length fatty acids, whereas mammals produce and metabolise even chain fatty acids. It was known that the breakdown products were propionic and acetic acids respectively and that while these two acids were similar chemically, acetic acid is produced in other human biochemical pathways but propionic acid is not. According to BP and ICI managers there was no *a priori* reason for thinking that propionic acid could have toxic effects on human metabolism, they therefore had not done any toxicological research on the metabolism of odd-length chain fatty acids. Their detractors could claim that until research was done no one could be sure of the safety of SCP, but by the time work had been done on fatty acids, both BP and Liquichimica had pulled out of SCP.

The environmental discharge of live organisms from the drying and harvesting processes into the atmosphere was a problem which really had existed in the Soviet Union; in Italy it was an alleged problem. ICI managers made the point that the ICI organism was so fragile that despite being emitted from the plant, unless it landed in a peat bog where anaerobic metabolism could take place, all the organisms would die. This was not necessarily the case with the BP and Liquichimica yeasts, but they controlled discharges. The question was raised whether the emission rates were still too high.[7]

BP used a yeast strain of Candida Lippolytica and Liquichimica a strain of Candida Maltosa. They both suffered from a confusion over the names, in particular with Candida Albicans, the yeast that causes thrush, which was certainly toxic and has the short-hand name of simple 'Candida'. There were then strains of each species of yeast which also differed in their metabolism and so in their possible toxic effects. One of the toxicological research managers thought that while all C. Lippolitica strains were non-toxic, some strains of C. Maltosa were toxic, but not Liquichimica's strain. He thought this had confused the Ministry of Health officials, especially as the Liquichimica process was based on the Kanegafuchi process which had been banned in Japan. He also thought that if this had caused them to have doubts about Liquichimica's strain so that they were afraid to license it, they would certainly not license BP.

> The trouble was that the scientists and bureaucrats in the Italian government were ignorant, not trained in this area, so that to say strain 1 is pathogenic and strain 23 is not did not convince them completely . . . especially when the debate amongst the experts was split. So there was a combination of political influence and no specific training to distinguish between strains of a genus

A decision was not helped by what one former BP consultant called confusing research generated by the soya lobby:

> There were some experiments in America, sponsored by the soya people. They took rats, irradiated them nearly to death, injected them with immuno-suppressant drugs, then injected several million

living yeast cells into the blood stream of these, these miserable beasts. After killing and dissecting them they found there was some persistence of living yeast cells, therefore the yeast was potentially pathogenic. Absolute, complete nonsense.

The Ministry of Health never did approve the SCP products, the Ministry of Industry therefore held back a licence to produce until Liquichimica went bankrupt in 1978 and BP scrapped their plant, adopting the official view that the soya lobby had been responsible for this disaster.

With the end of the Toprina project BP withdrew from all fermentation and most microbiological research. They were quite happy to abandon the fermentation technology they had built up because this was only a means to a commercial end for them:

> they were quite happy to leave it as a one-off development to see how it went. It was clearly stated that they were not in the business of developing a new fermentation industry.

Fishlock (1982) wrote that the BP board were in a state of emotional shock after the end of the Toprina project and that that was why they withdrew from the fermentation technology associated with Toprina. Others who have tried to talk to senior BP managers have come to a similar conclusion; for example, Sharp commented that he had succeeded in obtaining meetings with former board members and he believed the board was embarrassed by its involvement and reluctant to disclose their past thinking on Toprina.[8]

The only positive long-term benefit that BP gained from the Toprina project was the creation of BP Nutrition. BP Nutrition was originally part of BP Proteins, which was the division set up by BP International to manage the development of Toprina. BP Nutrition had the function of acquiring feed companies in order to give BP Proteins in-house nutrition expertise, but turned out to be profitable in its own right. When BP Proteins was run down after 1978, BP Nutrition remained and continued to grow and according to one BP manager, BP is now the second-largest animal feed company in the world.

It can be seen that the BP board controlled the SCP projects quite strictly and were involved in the detail of the decisions taken. The story of the projects' development is a story of the board's decisions, but the managers interviewed were interpreting and guessing the reasons for the decisions they made.

CONCLUSIONS

Relationship between R & D and the company

After reviewing the role of R & D in all these projects it is striking how the role and status of R & D varied between companies, seen from the perspective of those working on SCP projects. These views ranged from an exceptional view of the R & D department spear-heading the diversification of the company as in SugarCo, to the view that R & D was a mere holding operation which would only be turned to in a time of crisis. These attitudes were expressed by a narrow selection of personnel within the companies, but there is nevertheless a strong sense of the differing value that was placed on R & D activity (from the point of view of SCP workers). This could be summarised as in Box 5.1, which does no more than represent the opinion of a very select group and is most useful for identifying the way that they make sense of their organisational environment through the process of stereotyping; linking patterns of behaviour with identifiable struc-tures in the environment. This was also the case with the idea of 'culture', where groups within the organisations were associated with a particular perspective. This can also be represented as in Box 5.2. This is very much the manager-as-theoriser in action (see Chapter 1) and these boxes represent a shorthand way of making sense of a complex environment.

In addition to the variety of roles of R & D in the firms it is striking how varied were the organisational structures of the companies. No generalisation can be made here about how structures were related

Box 5.1 Value of R & D in different companies

BP/LQ/Shell	R & D exists in case company's core business deteriorates. Not essential to company success. R & D supports commercially based projects.
RHM	R & D activity unusual for a food company. Distinguishing and valuable feature for the company within the food industry.
ICI	R & D a vital activity that produced existing businesses and is expected to generate future business.
SugarCo	R & D moves to high-status activity with responsibility for diversifying away from core business of company.

Box 5.2 Culture or style linked to organisational sub-groups

Skilled-background culture	Chemical engineers
	Food scientists
	Microbiologists
Research styles	Understanding
	Empirical
Management styles	Food industry
	Chemical industry
	Oil industry
National culture	British firm
	Italian regulatory authorities

to project success, as external factors clearly played a large role in success or failure. Another feature of the content of this chapter was the frequency of crises external to the projects. The structure of the firm can be seen as enabling the process described in Chapter 1, where managers identify and negotiate an interpretation of such changes. Where there was an internal judgement that they were not effective in this role, they were changed.

From the perspective of the lifetime of these projects, the scale of incidental change was great with its succession of crises and project reappraisals. This encouraged managers in the process of adaptation of the project to circumstance. No project ended its life underpinned by the same set of assumptions with which it had begun. The process of enactment (see Chapter 1), of recognising relevant environmental change, the process of negotiating a consensus understanding of the meaning of this change for the project, and the establishment of a new agreed vision and course of action were all problematic and all essential to the process of management.

6 Technical choices in the development of the novel fermentation projects

As novel fermentation technology was developed there were key technical choices that faced the project managers. For dedicated SCP plant there was a set of criteria which had to be met for the plant to stand a chance of being economic, even with a favourable macro-economic regime. These included the achievement of a certain growth rate of the organism, minimum cell densities in the growth medium, the satisfaction of the biological oxygen demand within the fermenter with available technology and the attainment of high enough fermentation temperatures to allow low-cost cooling. The detail of the work in the R & D departments was in showing how the selected organism/substrate combination would meet these technical criteria if the process was scaled up. In the mycoprotein project one of the researchers described how prioritising research tasks was a matter for negotiation between research and the business side of ICI:

> I'll write programmes then people will criticise and say, well, that's a more pressing problem. So I then say to the business, well you define to me what the priorities should be and I'll research them. There's always arguments there because there are so many priorities to the business, that getting 1, 2, 3, 4 is quite difficult. And as soon as you think you've got one of them down, it just slips through your fingers . . . so it's never boring, never boring.

But such tasks had to be prioritised and dealt with even without such an explicit involvement of the business side of the company. The following sections examine the influences that led the projects to different technical solutions to their common problems of planning and building capital-intensive fermentation plant.

NOVEL FOOD RESEARCH AS A FASHION

Research on novel foods and SCP was scientifically fashionable from the late 1960s until the end of the 1970s. Part of the idea of 'fashion' in this scientific work was that the research was done simply because everyone else was doing it and because it was easy to obtain publishable results. An ex-BP SCP researcher and research director at RHM, thought that, 'Once SCP research became fashionable, and because it is a very easy subject in which to start research, a plethora of ideas appeared in the literature over the last 12 years or so' (Solomons 1983: 31). Solomons uses the idea of fashion without being highly critical of the SCP research field, but the more critical Moses and Rabin referred to a 'bandwagon effect' and, 'Once the bandwagon began to roll it gained so much momentum as to become unstoppable by rational arguments based on the unreplenishable nature of reserves of liquid fossil fuels' (Moses and Rabin 1982: 68). As a number of managers and scientists pointed out, this effect was not exclusive to SCP: 'there is a bandwagon effect in any area of science. If a university starts work and appears to be successful, then lots of people jump on the bandwagon'. This argument was just as true of some of the companies; the SugarCo fermentation project manager thought his company had benefited from working on what was generally perceived to be a socially useful technology.

> One major value for SugarCo – we did it when it was quite popular, quite a 'with it' thing to be doing. We did acquire a pretty good reputation as a result of this, and this undoubtedly helped the overall image of research within SugarCo. No question at all, on that basis alone, the money was well spent.

The pattern of activity within novel food project research also changed over time; an example of this was 'village technology' fermentation (see Appendix). A BP microbiologist told of the time when this was in fashion,

> There was this daftness of . . . there was an era where we went through village technology. You could chop up your locust beans, stick them in a fermenter, in Africa, make protein. If you look at some of the older SCP tomes, they talk about SCP from all sorts of biomass . . . [such as] bagasse . . . but you really can't run that sort of fermentation without really high technology. The stuff has to be made safe for the animals to eat, it's got problems.

As has been remarked before, it was the determination and success of BP in developing the technology that convinced many firms to start their own research. BP encouraged SCP conferences to promote the BP SCP projects and SCP generally and the conferences and the industrial effort in SCP research made the field attractive to hundreds of university and government research institutes around the world. In these conferences a whole set of potential projects were being discussed:

I think that's right, it wasn't just SCP, it was a whole lot of other microbial processes, at the same time SCP people were talking about using polymers for tertiary oil recovery by fermentation . . . they were talking about updating processes . . . like citric acid [fermentation] and about going back to processes which historically had been used in the Second World War to produce alcohol, etc. . . . so it was a whole combination of things that were happening. But at the heart of it was a wish to develop sophisticated fermentation technology.

For a time it appeared that SCP would be *the* sophisticated fermentation technology that would be developed. The technology was intrinsically exciting in the way it offered to replace humankind's fundamental dependency on agricultural production systems and in particular the agricultural nitrogen cycle. As a research scientist said,

It's simple and it's basic. You've got your nitrogen in the atmosphere, you've got routes by which that is fixed and in the soil it is oxidised to nitrate, and that is well known. But the important step is biosynthesis, in a green plant, to make plant protein . . . that is what man wants. You are absolutely tied to that bit which is the green plant growing in the field taking inorganic nitrogen from the soil and making protein.

SCP technology would give humankind the ability to bypass the green plant protein synthesis step for the first time on a large scale because the SCP organisms synthesised protein from simple nitrogen compounds such as ammonia or urea, provided they were supplied with a chemical energy source to replace the sunlight that a green plant uses.

This 'science fiction' image of SCP combined with the perceived need to feed starving millions in the near future was the source of many scientists' enthusiasm for the SCP and novel food projects, and hence a partial explanation for the bandwagon effect. This remains true today, with Villasen writing of the enthusiasm of the scientific researchers, on the Danish Bioprotein Project: 'It has probably also been important that the project with its quite obvious social perspectives has inspired scientific researchers at the University of Odense' (Dansk Bioprotein 1988a). Villasen goes on to link this enthusiasm on social-benefit grounds to what he believes to be the astonishingly rapid progress of the project.

But by 1984 the bandwagon had gone into a ditch and there is a severe drop in the number of references to SCP in the literature. According to Fishlock the Pruteen project accounted for over a third of all ICI spending on 'biotechnology' by the early 1980s (Fishlock 1982), and Antebi and Fishlock refer to the ICI venture as the largest investment made by a single firm in biotechnology research (Antebi and Fishlock 1986). Despite the historic importance of SCP in the development of industrial biotechnology, many general biotechnology texts fail to refer to the technology at all after 1984; Silver (1986), Knorr and Sinskey (1986) and Olson (1987) are examples. One mentions SCP, but only to say that it 'obviously makes good sense' since SCP provides an alternative to dependence on green-plant protein synthesis (Prentis 1984). Another refers to the RHM work, but discusses it as a further development of SCP technology and describes the product as a 'fungal SCP' (Daly 1985: 14).[1] Daly's comments tend to confirm that SCP and Mycoprotein have been perceived as scientifically and technologically closely associated and that it has only been Marlow Foods that has been interested in breaking the association for marketing purposes.

AN INFORMAL SCIENTIFIC NETWORK

The common skill base of the SCP projects, the long duration of the projects and the bunching of the project start dates allowed an informal network of relationships to flourish among the research scientists working on different companies' projects. This network was described by one manager as

> a group of people in the UK who had been working on areas highly relevant to SCP production and they all knew each other from the academic community. Then all of a sudden, they went to work, over a period of, 3, 4, 5 years in the Shells, the ICIs the BPs of this world. Clearly they still knew each other. So there was interaction between them, because they still went to the same scientific meetings.

A departmental manager commented that

> Technologically its a small world. I knew the people in Shell . . . I worked with the people in ICI and so on . . . not on those

particular projects necessarily. Yes, with Hoechst I had specific dealings with them, or some of them . . . all the people in the UK that have been involved in it [SCP], know all the others . . . we went to the same universities.

The personal network of contacts had its basis in shared scientific interests and training and this remained in place when formal titles and organisational forms changed. When talking about the personnel changes that occurred at the end of the Pruteen project, one manager commented that

that's rationalisation. What companies call themselves and how they appear to the world is one thing, the people are the same and they have to a large extent remained the same in both John Brown and ICI . . . I'm still talking to the same people, but our job titles change regularly – it doesn't mean much.

SELECTION OF ORGANISM

The organism screening process and the selection criteria

The screening process was the important first step for all projects, one manager describing the characteristics of the organism as, 'probably 90 per cent responsible for dictating what the product was going to be . . . but there were ways of manipulating growth to just tweak up a few other things'. A summary list of the key organism selection criteria would be the following (abstracted from researcher comments and a paper by Hamer *et al.* 1976: 67):

Should not produce toxins or be pathogenic
High yield, e.g. in grams of product/grams of material consumed
High concentration of organisms in medium, e.g. in grams/litre
High growth rate, e.g. as cell mass doubling time
High productivity, e.g. in grams/litre/hour
High stability of the culture
Resistant to contamination
Thermo-tolerant
A low percentage of cell mass that is nucleic acids
A low percentage of cell mass that is mineral salts (ash)

Small variations in organism characteristics were expected to have a large impact on profits, so the selection process was more rigorous than any previously performed for industrial purposes. Most companies allowed several years for organism selection and in this time the selection techniques themselves evolved, as will be discussed below. The company would fix a set of selection criteria that it believed most suited its commercial objectives and then trawl through the many possible organisms to draw up a short list of candidates. The selection criteria could then be further tightened to whittle the candidate organisms down to one. One of the RHM managers described this process in his company's case:

> When we decided that what we needed was texture and that was synonymous with some member of the mushroom family, we identified up to 3,000 organisms world wide as possible candidates to provide the combination of structure, texture and high growth [we wanted] . . . Out of that initial characterisation came a handful of organisms that seemed to meet the criteria . . . and then only two of that handful. One was a penicillin, which are probably the best understood fibrous organisms. [But] after three or four years of research there just wasn't any way that it was going to give us the high yield growth characteristics that we were interested in.[2]

An organism was selected by criteria which were related to its use in large-scale continuous fermentation, but how it would really function under such a fermentation regime could only be found out by small-scale pilot fermentations. This stage of the research would take at least another one to two years – and in RHM's case led to the rejection of the original organism and the loss of several years's work. The problem was described by one of the ICI research managers:

> It's a bit of a chicken-and-egg situation, because you select the organism on criteria which are based on the desired product specification . . . [then] you learn more about it at the pilot plant stage and so on. You get to a stage where you are committed to the organism which you have chosen relatively early in that development process . . . it's necessary to choose early because of the need to do all that toxicological work . . . you are testing organisms for characteristics which you think will be important in the large-scale process but you need to know the characteristics of the large-scale process to do that. It's based very much on how you conceptualise the process at an early stage.

When product conceptions changed, so did the criteria for a 'best' organism, and this happened when the RHM project was switched between a Third World nutrition product and a First World moist food. The RHM head of research in the 1970s talked of the advantageous amino acid profiles of fungi compared to those of yeast and bacteria (Spicer 1971), a difference only critical while the importance of protein nutrition is accepted. In a First World food market context the marginally advantageous amino acid balance was not an important issue for RHM, but the texture of the product was.

The RHM organism is now grown on the Pruteen pilot plant but this was not designed to grow a filamentous fungus for use as human food. ICI and RHM are investigating how changing the fermentation parameters might change the product characteristics. One of the researchers involved described the situation:

> optimal growth for the organism might not give you the raw material at the end of it that is optimal for food quality. And this is what we don't know at the moment, that is what all the research is going in to, how to characterise it, how to make it optimal.

SugarCo and the smaller waste-SCP companies did not screen so many organisms as ICI, BP, RHM, Hoechst and IFP, and the selection of organisms was more *ad hoc*. SugarCo chose a fungus because fungi can grow on a range of substrates, could be harvested by simple filtration and had relatively low RNA content. Fungi also tolerate acid pH, something that bacteria are sensitive to, and so SugarCo could use an acid fermentation as an approximation to sterility. This did not protect their Belize pilot plant against infection with a competing yeast with an optimal growth temperature of over 46°C. This organism successfully competed with the SugarCo fungus, Aspergillus Niger, and this 'technical' problem forced SugarCo to move to a solid-state fermentation rather than a liquid fermentation. Such a contaminating infection was a fear for all the projects.

For an SCP process to have a chance of being economic many of the 'accepted' and established values of the selection criteria listed above had to be radically improved. One of the Shell research managers gave an example:

> a lot of the research went on getting high productivity cultures remember, very few people had grown organisms to any sensible density at that time. Most people regarded 1 or 2 grams/litre as a high density of organism. Before you could ever think of being economic you had to have 20–30 grams/litre, an order of magnitude greater.

High cell densities were important because of the huge energy cost of dewatering and drying – the higher the cell density of the fluid emerging from the fermenter, the lower dewatering (by centrifuging) costs would be.

One of the Shell researchers thought there was an overemphasis on yield coefficients in the literature, as a means of comparing the relative worth of different organisms, because yield coefficients varied for the same organism under different conditions. For Shell, with their methane-based processes, the problem was the waste of feedstock, not the percentage of feedstock that was utilised that was then converted to biomass.

> It's very hard to get a high feedstock conversion . . . everyone tells you what the yield coefficient is, how many kilos of product for how many kilos of substrate utilised. But substrate or feedstock supplies, the story is very different. In fact you have to achieve something like 80 per cent conversion of the methane . . . most processes will only give you 30–40 per cent, some considerably less than that. A serious problem.

This researcher pointed out that this was the major block to all the algal, hydrogen-based and methane-based SCP processes, since they all relied on gaseous substrates, (carbon dioxide in the algal cases). For Shell and Dansk Bioprotein, at 80 per cent conversion of substrate to cell mass it was possible to enrich the wasted gas and burn it in the spray drier, or to dry it, remove carbon dioxide and recycle it to the fermenter. Shell began researching the problem of engineering the recycling of the methane–air mixture, but,

> Recycling the methane for reuse to avoid the low utilisation rates cost money, and you compressed a mixture of methane and oxygen. You can do this, but there is the possibility of explosion . . . [then] you have to extract the CO_2, need to dry it . . . yes, you can, we worked on methane recycling – obviously, you can do it easily on the lab scale, but on the scale of a commercial plant . . . and all these costs

The Dansk Bioprotein company literature claims that 'Less than 20 per cent of the natural gas injected is released to the environment' (Dansk Bioprotein 1988a: 19). Rather than working on recycling, Dansk Bioprotein suggest that they might flare the gas off or arrange for it to be burnt in an adjoining company's plant, thereby replacing the engineering problem of how to recycle an explosive mixture of gases with the partly organisational problem of finding another company to take the waste natural gas. They have not yet decided what to do with this waste gas and Shell thought this problem so acute, and the methane-to-methanol step so difficult for the methane-consuming organisms, that, 'at the end of the day, ICI probably had the best hydrocarbon-orientated route'.

The search for thermo-tolerant organisms

Another major aim of research was to find thermo-tolerant organisms because a success here would have enabled the projected capital investment to be lower because there would be less need for cooling the fermenters. The rate of growth of the organism would be a maximum only for some optimal temperature which suited the enzymes that made up the biochemical pathways of its metabolism. It was important that growth rate was maximised, so this optimal temperature had to be maintained by cooling equipment that removed the heat of fermentation and prevented the fermenter temperature rising, thereby slowing the rate of growth. But if this temperature was near the environmental temperature, the cooling equipment would have to transfer much larger quantities of heat out of the fermenter using heat pumps – an operation that added to capital and operating costs. If the organism had a naturally high optimal growing temperature – in the high 40°C – these costs could be lowered because the greater temperature difference between the fermenter and the environment would lead to a naturally higher rate of transfer of heat out of the fermenter.

So the BP organism had an optimal growth temperature of 30°C and this was so little different from room temperature that the costs of cooling were relatively large and a question hung over how the plant would operate in typical Mediterranean temperatures of 30°C. BP actually found a strain which would grow at a higher temperature but which they could not use because the plant was already being constructed. According to a member of the project,

> The economics weren't critical, but it was at the stage where we were honing the process and were interested in 1 per cent improvements in production. That was a margin that could contribute to profits.

The Dansk Bioprotein organism has an optimum growth temperature of 46°C, high by the standard of the researchers of the 1970s. The Dansk process still requires cooling, but less than, say, the BP organism required.

Most organisms have an optimal growth temperature in the range 15–45°C and one of the spurs to research for thermo-tolerant organisms in the 1970s was the claim by Phillips Petroleum to have found a thermo-tolerant organism with an optimal growth temperature above 50°C. One of the Shell researchers, now Professor Hamer at Zurich Polytechnic, recalled that there was something of a technical debate at the time over whether these thermo-tolerant organisms could be used effectively.

> Everyone used to say, oh, well, the solubility of methane or oxygen goes down with increasing temperature . . . but of course if you put pressure on the system it doesn't make that much difference. What everyone also forgot, was the diffusivity of gases in liquids will increase with temperature, and this almost compensates for the loss of solubility.

Nevertheless, the effort put into the search for such organisms was large.

> ICI put in, a 100 man-years maybe, I suppose Shell put in 20, 30, 40 man-years of effort looking for these organisms. Everyone said, this is a false claim in the patent literature by Phillips. It wasn't a false claim, I can tell you that. I had at least 10 or 12 methanol utilisers that grew well at 50–55°C here [i.e. Zurich Polytechnic].

Hamer believed that the search techniques in general use at that time were not capable of showing up the thermo-tolerant organisms, although Phillips Petroleum must have been using a more innovative procedure which they didn't reveal in the patents for their SCP process. By the end of the 1980s thermo-tolerant organisms are routinely encountered in culture collections.

The choice of an aseptic fermentation process

An aseptic fermentation is one which is resistant to infection by foreign organisms. The novel food production processes had to be reliably free of pathogenic organisms to produce a sterile product and there were a number of approaches to obtaining a consistent and safe product:

1 Fungi tolerate acid pH but bacteria do not. Since most pathogens are bacteria the choice of a fungus in an acidic medium would tend to exclude the possibility of bacterial infection. The fermentation could then be open to the environment, as was the SugarCo process and the BP gas-oil process.
2 Choosing a substrate upon which micro-organisms grew with

difficulty, making infection unlikely. The Shell and Dansk Bio-protein processes based on methane are the principal examples of this.

3 Engineering to sterile standards. BP and ICI used this method to guarantee that no foreign organisms could enter their fermenters.

Toxicological monitoring of the product was a part of all the processes, but especially important within the first two options. Sterile engineering is expensive and Dansk Bioprotein claim that the non-sterile nature of their process is one of the key features that enables them to expect reduced full-scale plant construction costs, only 10 per cent of the cost of 'earlier designs', by which they mean the ICI Billingham plant[3] (Dansk Bioprotein 1988b: 9). A member of the John Brown Pruteen construction team explained that

> Essentially ICI could see that they were going to get very high carbon conversion, but they could only do that with this particular organism, and it was sensitive to infections. Sterility was not the requirement of the product, but of the organism. The high dependence on feedstock cost made it important to have high conversion efficiency.

The problem of infection when working with bacterial cultures affected the IFP choice of organism, who worked for a time with both yeast and bacteria:

> and we finally chose yeast. We continued to keep our view that if we worked with bacteria we would be obliged to work in sterile conditions; with yeast you worked with, clean, but not truly sterile conditions. It would have been easy to keep this [bacteria-based] laboratory process sterile, but not a plant or even a pilot plant.

There was another reason for choosing the most rigorously sterile process. This was the need to convince sceptics that the novel product was completely safe. Among microbiologists there was some doubt that a process that relied on routes 1 or 2 above was beyond suspicion of infection. There was always the chance that some organism previously unknown and rare would find its way into the fermenter and survive to contaminate output, where safety would rely on the detection of contamination by the firm. Sterilisation of inputs to the fermenter combined with sterile engineering would remove all doubt about contamination and it was a guarantee of the absence of contaminating infections. The issue of sterility was especially sensitive for ICI because they had seen how minor scientific doubts had been exploited to BP's disadvantage in Italy.

Riviere has described how the BP Lavera process used control of pH, temperature and dilution rate to control sterility (Riviere 1977) while one of those who worked on the plant thought that the importance of high standards of sterility was overemphasised by some bacteriologists and microbiologists.

> You see it's psychological . . . we did it [septic fermentation] in Lavera and it worked! At the very beginning I enquired about contaminants at the pilot plant. I decided to see if the bacterial contaminants were toxic and if it [the yeast culture] was stable or not. It was innocuous and the bacterial population was absolutely stable. So why bother with asepsis?

This same researcher suggested why the scientific community continued to have doubts on non-sterile biomass fermentation:

> Biomass production is a unique problem. Usually companies producing metabolic products, biotechnological processes, optimise their strain by genetic engineering, mutation and so on, for them it is essential to be aseptic. The only exception is when you want to produce biomass.

Such a genetically altered organism would likely be susceptible to infection from the environment, so sterility was a necessity, although organisms taken from the environment might not need sterile conditions to be resistant to infection. The common industrial experience was of a need for asepsis and so there was a generally critical attitude among the microbiological community towards non-aseptic SCP processes.

A mixed or pure culture?

Another division among scientists at the time which was linked to the need for sterility was the question over whether a pure culture fermentation was better than a mixed culture fermentation. One of the Shell scientists commented that

> One of the big debates of those days was whether a monoculture was necessary or not. A supposed danger of mutations. If the process was either sterile or with one extra culture, there was a high certainty of the output.

The suspicion of mixed cultures was that perhaps genetic material could be exchanged between the separate strains or that over time the relative proportions of one strain to another would change as mutations occurred. It could not be proved in advance of lengthy experience with continuous fermentation that such things would never happen, so the debate could not be resolved. According to the head of R & D at Dansk Bioprotein, scientific suspicion of mixed cultures persists today.

But a strong argument for the case that mutations were not a problem was that if the conditions for production were strict there is no reason for a mutation to survive. If there is one it will be maladapted to the controlled environment and so will die. The BP Lavera plant used a non-sterile mixed culture and one of the projects scientists remembered a conversation he had once had with an eminent bacteriologist:

> the father of penicillin Ernest Chaim told me that any industrial process using bacteriology should have a pure culture. I asked him why, he said it is obvious, but I told him you are denying the existence of cheese. In my country peasants produce Camembert every day and it never turned out that the next morning they have Cheddar. Cheese shows that a mixed culture, when very carefully set up, is absolutely safe . . . you know, bacteriologists spend all their time with test-tubes and have no imagination

ICI, RHM, Hoechst, IFP and BP at Sarroch used pure cultures. The principal mixed culture processes were the methane-based projects of Shell and Dansk Bioprotein. Shell tried to use a pure culture but was forced to the conclusion that mixed cultures were more effective and that there was no problem with their stability.

> We made a number of fairly basic discoveries and one of the important ones was that . . . you did indeed get cultures that utilised methane, but as you purified them more and more, they would reach a point where they became much less active . . . it proved to be very difficult to get any sort of pure culture of a bacterium that would use methane that would grow at any sort of presentable rate. But just before you got to that point you got a mixed culture that grew extremely well.

Shell found that the reason for this was that the methane-utilising bacteria produced methanol as a first step and that methanol acted as a toxin which inhibited this first organism's growth. So a second organism was necessary to mop up the methanol and allow rapid, continuous growth of the first. But small quantities of toxic formic acid and formaldehyde were produced by these organisms, and if other organisms were present in the culture that specialised in the oxidation of these products, the efficiency of the culture could be enhanced still further. Since natural gas is not a pure substrate and gas from different fields contains different proportions of the gaseous alkanes, a range of acids and aldehydes is produced depending on whether methane, ethane or propane were being digested. Again, mixed cultures would mop up a range of intermediate products, and Shell isolated several synergistic mixed cultures of this sort,[4] while the Dansk Bioprotein mixed culture is another example. Shell were investigating these bacteria without hope of developing an SCP process.

> More or less up to that point we had been thinking of taking methane and making some attractive chemicals, rather than SCP, the reason being that if you looked at the growth rates that were reported in the literature, they were pathetic, the yields were very, very low. It didn't seem that SCP production was at all realistic. But when we discovered this interactive thing . . . we realised we were getting growth rates in the right ball park to make SCP a possibility.

The other companies were using substrates with a wealth of possible organisms and so could choose a pure culture, a choice which avoided some of the doubts about the reliability of such processes. ICI had originally begun work on natural gas substrates in 1967, but switched to methanol a year later in the face of the problems of low productivity, and the recycling of explosive methane–air mixtures. By using methanol as substrate they replaced an inefficient biological step with an efficient chemical step and gave themselves a larger pool of candidate organisms and a feasible pure culture process – but they had the option of chemically produced methanol on site which was unavailable to Shell. It remains the case today that on a natural gas substrate a mixed culture offers the only possible economic route to SCP production.

FERMENTATION CHOICES

The low profit margins on the SCP products required that continuous fermentation be achieved. 'Continuous' fermentation meant longer fermentation runs than were at that time common in the brewing and pharmaceutical industries and a less labour-intensive operation then was associated with small-batch production. With the prevailing mode of batch fermentation there was a widespread belief that continuous fermentation was not possible, one manager commenting that this was where a lot of woolly thinking came in, and the pharmaceutical companies had a lot to answer for – in forming attitudes. First to continuous fermentation, then to continuous aseptic fermentation.

The pharmaceutical companies typically fermented high-value products in batches and tended to doubt the possibility of long fermentation runs, simply because they did not do it. They were even more incredulous at the idea of long, aseptic fermentation runs. Their attitude was important because they at least had fermentation production experience – BP and ICI Agricultural Division did not. One of the RHM managers reinforced the idea that continuous fermentation was seen as infeasible at that time, when talking of the RHM R & D team's achievement:

> they applied some frontier-breaking, front-rank science. Had you asked anyone in 1970, can you grow anything in a continuous culture, everybody said no. Nobel Prize winners said forget it, stupid idea that. We did it.

As well as the operational requirement of continuous fermentation, the companies had to choose an appropriate fermenter design. The major choice appeared to be between the traditional stirred tank or an air-lift fermenter. One of the Shell project managers described Shell's comparison of the two types of fermenter:

> A major engineering study for scale was performed and there was an oxygen delivering problem . . . the economics came out for a stirred tank – what they've been using in brewing for hundreds of years, but it has much greater flexibility than ICI's. If we had scaled up we would have had a dedicated plant with a stirred tank.

This problem of delivering sufficient oxygen to all parts of the fermenter fluid grows with the size of the fermenter. If air is injected through some form of nozzle assembly it still has to be mixed efficiently with the fluid so that all parts of the fermenter volume are oxygenated. In a stirred tank some kind of paddle device would ensure that the fluid was flowing and distributing dissolved oxygen to all parts of the fermenter, but an air-lift fermenter relied on the incoming injected air to stir the liquid and to supply sufficient oxygen for maximum growth rates. On a very large scale Shell found that the stir-tank fermenter nicely solved the oxygen distribution problem.

BP also chose the stirred tank for the gas-oil and normal alkane fermentations. BP were proud of their fermentation achievements, especially with regard to ICI, who were usually referred to as having the most sophisticated fermenter technology. One BP manager commented that, compared to ICI, the BP Sarroch plant had,

> A totally different fermentation style, but BP were doing massive-scale fermentation five, six, seven years before ICI, 2,000-hour runs. Totally uncontaminated, monoseptic, truly continuous. The fermenter was based on the traditional Rushton stirred tank, S-bars, agitator driven. The old equipment brought up to the highest levels of aseptic operation by BP engineers.

And another commented that 'Personally I get a bit sniffy when people go on about the ICI fermenter'. This was because the total fermenter capacity at Sarroch was greater than that of ICI's single fermenter. Each of the three Sarroch fermenters was of 1,000 cubic metres capacity, each was served by a compressor, equivalent in power to a Boeing 707 engine. This arrangement delivered a safe protein product and met the need for continuous aseptic fermentation. What bothered the manager quoted above was whether the advanced technology of the ICI fermenter was really necessary as part of the process to produce a safe and economic SCP product.

Liquichimica chose the same type of stir-tank fermenter as BP but with the difference that they had two fermentation stages, so that the yeast was allowed to continue to ferment in a second fermenter without the addition of further n-alkanes. This meant the substrate could be more effectively converted to cell mass and ought to have been to their advantage in the toxicological debate in Italy since their protein product contained less undigested alkane substrate. The Health Authorities did make official depositions noting that Liquipron contained less residual n-alkane than Toprina when used as a feed.

Dansk Bioprotein have also chosen to use stir-tank fermenters in their natural-gas-based SCP process, but they plan to use a series of ordinary 50 cubic metres stir-tank fermenters, rather than attempting to scale up to the capacity of the BP, ICI or planned Shell fermenters. These 50 cubic metres fermenters can be bought from equipment suppliers and require no specialised engineering whereas the large-scale fermenters have to be designed and built to order.

The ICI fermenter was the most complex of those used for SCP fermentation and was basically an air-lift fermenter, which ICI sometimes called a pressure-cycle fermenter. A manager in the John Brown design team described this object as

> 60 metres high, the largest continuous fermenter in the world . . . it still is. So we are talking of the sort of technology that verges on atomic reactor technology in terms of the amount of metal and

the accuracy with which it had to be put together. Monoseptic operation with guaranteed continuous fermentation for a year at a time. What that meant was that nowhere could there be a hole of equivalent diameter to 1 micron [a millionth of a metre]. The sterile envelope didn't just consist of the fermenter, which is large and complex enough . . . the metal that went inside there weighed 600 tonnes, because there was something like 100 kilometres of pipe in there to inject the methanol, in about 20,000 different places. All internal to the fermenter envelope.

The commissioning manager for the Pruteen plant gave the reason for building a large fermenter of 1000 cubic metres capacity:

It was 75,000 t/a, the thing that dictated that was, one, the relationship between scale and cost-effectiveness. Once [past a certain scale], you are not going to get any more cost benefit from it, no matter how big you build the plant. So ICI took a point on that curve [price per unit production versus fermenter capacity] which was dictated by the fact that we'd run a pilot plant for many years at 45 m^3 and we wanted to go up to 1,500 cubic metres and that scale-up factor was the most comfortable that people were willing to take a risk on.

A manager in the John Brown team described how the fermenter was designed:

They [ICI] had in their midst quite a brilliant mathematician, who was able to devise a mathematical model for the ideal way in which a micro-organism could thrive on methanol, which is toxic at any greater than very small concentrations. So what was necessary, they soon found out in the laboratory, was to feed a micro-organism with a carbon limitation in order to achieve a high carbon conversion efficiency at very low concentration of the substrate, but with no restraints with regard to oxygen or heat transfer. Now that was a tall order and it really meant a new approach to fermenter design. And this fellow devised essentially the prototype of the pressure-cycle fermenter, in which air is injected at the base and because of the very high head, there is a very high driving force to dissolve oxygen into the solution. Provided there is good mixing and it is present in sufficient concentration that organism can take up the oxygen very rapidly. It's getting the oxygen in that is the problem, always has been with fermenters.

The fermenter consisted of essentially two columns of liquid (see Figure 2.1),

> A riser column of a mixture of gas and liquid and a descender column. Providing you keep on injecting air at the riser base then the power you put in is quite efficiently converted to circulating movement in the fermenter, which provides all of the power for oxygen transfer and mixing. So mixing was the other problem.

The mixing problem required more exceptional mathematics:

> This mathematician devised a modular concept in which the liquid would have to go through orifices such that larger bubbles would be broken up without forming very fine bubbles that would be difficult to disengage at the top. At the same time there was a point just below each orifice at which methanol could be injected and efficiently mixed.

The bubbles were therefore constrained to the optimal size for gas exchange to the liquid and for disengagement at the top of the riser. As the liquid–gas mixture approached the top, the width of the riser increased so that there was 'an enlarged area with baffles that I won't even attempt to describe because it takes too long'. In this area the liquid flowed more slowly, the shear forces in the liquid decreased and the bubbles coalesced and finally disengaged from the liquid. Greenshields and Rothman remark that foaming was a problem in the ICI fermenter (Greenshields and Rothman 1986). This would have resulted from the bubbles of gas not having coalesced sufficiently by the time they reached the top of the fermenter, so that they did not disengage from the liquid. The orifices which were designed to constrain bubble sizes were probably redesigned to solve the foaming problem at the same time as the methanol-mixing problem was solved. An academic fermentation consultant to BP and ICI was critical of these technical problems.

> A major and very technical clanger . . . early on in the ICI scale-up . . . it could have been anticipated from the small-scale experiments, but it wasn't and it resulted in lots of work which normally you would have expected to have been done at the beginning. Now it had to be done halfway through, many modifications to the fermenter were necessary, and it was already built. Ultimately, these were scientific oversights.

The main problem was a poor mixing of methanol and water, which slowed the rate of fermentation in certain parts of the fermenter. One manager commented on the technical solution:

I think it would be fair to say . . . that they went from 1,000 to 20,000 nozzles . . . from a large number to an even larger number . . . and that involved a lot of detailed work of cost optimisation; the cost optimisation comes in what techniques do you use for fabricating a nozzle, techniques for siting a nozzle and piping up a nozzle. And how to cut the holes and baffles . . . we looked at a whole range of fabrication technologies and put them together in a way that hadn't been done before.

There was more novel technology in the design of the sterilising equipment (for sterilising the fermenter inputs), which was engineered to higher sterile standards than any achieved before.

When we first got involved [John Brown], we knew that we needed kill ratios of 10 to the power of 23. In sterilisers you can't in fact totally and absolutely kill all infections coming through, but what you can do is to design for a ratio of how many come in and how many get through. And our ratio was about a million times better than anyone else normally attempts.

The complexity of the technology added to the reasons for building big.

It was recognised that there would be an economy of scale with such fermenters and in order to produce SCP in quantities to satisfy the feed market, fermenters would have to be quite large. That would result in high capital investment and a demand that the plant should be kept running for very long campaigns.

The energy consumption was huge and added greatly to costs when the price of hydrocarbons rose in the late 1970s.

We had a 10 MW compressor pushing air into this fermenter. What we were doing was imparting the momentum equivalent to a Concorde moving supersonically . . . the same amount of momentum was in there . . . over 1,000 tonnes of liquid moving at 3 m/s.

The technical problems were overcome in time for the plant to be commissioned within three years, which was only six months longer than originally planned, a 'fast track' performance considering the novelty of the technology involved.

HARVESTING THE BIOMASS

Harvesting was the process of separating the cells from the fermenter liquor. For the yeast and fungal processes this was relatively simple since yeast cells were large enough to centrifuge out of solution, after which they could be spray-dried ready for packing and sale.[5] In the late 1970s and early 1980s centrifuging technology was unable to separate the much smaller bacterial cells from their growth medium, so the two companies with bacterial processes, ICI and Hoechst, had to devise new ways to harvest bacterial cells. The ICI process administered a temperature and pH shock to the cells.

> You can't just put them in a centrifuge or a filter, nothing will happen . . . they are about a micron and a half long and half a micron wide . . . so you've got to make flocs, which are perhaps 0.1 mm. The idea was that the temperature shock would open up the cells to some extent . . . and the pH shock would change the isoelectric point whereby the cell proteins would tend to coagulate . . . and the addition of acid and depressurising . . . meant a lot of the CO_2 would come out in the form of very fine bubbles together with the populated biomass, which would form a float.

This float of coagulated bacterial proteins was scraped off the surface of the growth medium in a sterile fermentation tank, degassed, centrifuged and then dried. It was this treatment of the bacterial proteins with acid, and then heat in the drying stage which rendered them almost insoluble in water. This property was useful when the proteins were to be animal feed, when there was a need for feed to resist rain and wetting, but it rendered them useless as a human food protein. When ICI began working on human food applications of Pruteen, a major aim was to render the proteins soluble in water once more by hydrolysing the Pruteen proteins with enzymes.

Hoechst were trying to develop human food proteins from their bacterial fermentation process from the beginning of their project and so tackled harvesting in a way intended to avoid denaturing the proteins and to preserve their solubility. But if the protein was intended to be used as human food then the bacterial nucleic acids still had to be removed. Hoechst patented the 'Schlingmann process' as a harvesting method, named after one of their SCP research project managers. One of the other research scientists on this project described the Schlingmann process:

> This process he invented, very simple, much, much better than any other kind of way of destroying the nucleic acids that other patents did. He treated the cells with a methanol–ammonia solution, this opened the cells. The nucleic acids were soluble in the methanol solution, the proteins were not, so the nucleic acids

could be washed out . . . We had 90 per cent total protein after, compared to 70 per cent before . . . less than 1 per cent nucleic acids.

ICI and Hoechst had had a general technology exchange agreement since the early 1970s where they undertook to discuss the exchange of technology to their mutual advantage and they began discussions on the use of the Schlingmann process in the production of Pruteen. Hoechst thought this was a unique piece of downstream processing technology which they could offer to ICI in exchange for biomass from ICI's Pruteen fermenter. They thought that cooperation also made sense because both companies had processes based on methanol that used similar micro-organisms. However, negotiations were first delayed, one of the Hoechst research scientists attributing this to problems of communication between higher levels of management in the two companies, then failed completely when ICI found that UK food safety regulations made success more likely if Pruteen was processed in the UK rather than in Germany. Subsequently Hoechst approached SugarCo and RHM to discuss the use of the Schlingmann process, but cooperation never progressed further than a few informal discussions.

By the time Dansk Bioprotein came to design their SCP process, modern centrifuges were available that could harvest bacterial biomass directly without the need for a flocculating step or the Schlingmann process, the developments having been made by the centrifuge manufacturing companies as a response to anticipated biotechnology industry needs. The centrifuges give a 10 per cent dry-matter concentrate which is then filtered to give a 30 per cent dry-matter suspension, which is heat treated and then spray dried. The Dansk Bioprotein spray driers incorporate scrubbers to minimise the release of microbial dust to the environment, another equipment manufacturer innovation; BP engineers had to design and build their own scrubbing equipment. There were other equipment innovations, such as the availability of sensitive oxygen concentration sensors which led the Dansk Bioprotein R & D manager to describe the equipment manufacturers as having adapted to a 'general trend' in demand, the perception of this trend having been established partly through the activity of the SCP ventures. The importance of these equipment innovations is that they helped to enable the Dansk Bioprotein venture – this small company did not have the in-house engineering skills to tackle the design of such equipment.

TOXICOLOGY

The companies that built full-scale plant wanted to be sure their products would be considered safe when they were ready to produce them. They had two major problems:

1 There were no standard government procedures for the testing of novel foods.
2 The microbial nature of the foods led to greater doubts of the safety of these foods compared to agricultural sources of food.

Test procedures needed to be worked out and agreed upon, preferably in cooperation with government bodies. This made the toxicity testing of the products at once difficult and very important and far tougher than a new agricultural product would have to contend with. One of the managers of the RHM project put the cost and time that RHM was involved in toxicological and nutrition tests at £20 million over 13 years and one of the ICI managers put the cost of toxicity trials to ICI at a third of the total research cost of the Pruteen project.

Some writers have observed that when the EC Commission finally published its SCP testing regime it restricted the levels of copper and zinc allowed in SCP yeast feeds, but made no such restriction for animal feeds in general (Sherwood 1974). Another general observation of interviewees was that it was common for soya and fish feeds to contain aflatoxins and levels of bacterial infection unacceptable in SCP feeds, which were effectively sterile. ICI and BP were not simply negotiating test procedures in a neutral environment, they had to show that the assumptions behind tests for drugs were inappropriate for novel foods. An ICI manager remarked that ICI

> 'did more for the animal food tests than are done for most human foods you eat. We were just thorough. The regulatory people around the world did not know how to test Pruteen. Most drugs, you look for a minimum of residue in tissues, but with SCP you want a maximum.'

The design of toxicity tests was critical given the absence of test protocols.

> It all boiled down to this, that with a novel product you must look at it carefully before testing. If it differs significantly from conventional foods you cannot test it conventionally. You will get poor results, not due to the material's quality but due to incompetence in designing the experiment.

Test design could affect judgements as to the safety and toxicity of the product. An example of this was the potassium (K»fH) deficiency of Pruteen.

> For example . . . there is no K»fH in Pruteen, but conventional feeds contain 1 per cent of K»fH. We need a base [alkali] in production, but the cost difference between sodium and potassium hydroxide is so huge that we use sodium hydroxide. This doesn't matter at all provided the mineral balance of the diet is adjusted to have the same concentrations of both minerals. If no K»fH is added to Pruteen it will fail to perform satisfactorily in feeding trials. This is very specific knowledge, but we convey it to potential purchasers and these precautions are normally taken.

The RHM testing procedure was even more rigorous than for the novel animal feeds:

> it's easy with animals. You just give it to them and see what happens . . . Before you could get permission to test on experimental volunteers you had to show a much wider range of tests running through lifespan studies to many species, multi-generation tests, across a number of different species. At the time, not required anywhere else, but something we had involved ourselves in, was the whole teratological testing on the unborn. A series of tests that tried to chart the equivalent biological value of the food, in terms of the diet that a human being would have, not just a bulk building diet you might give to an animal . . . it's all very standard now . . . one of the things that came out of this project was to help the government define its set of tests for new and novel foods. Much of the Committee on Novel and Irradiated Foods has as its genesis the mycoprotein project. That is one other big benefit that I think has come out of this project for UK Ltd.

The benefit to RHM of this attitude to tests was a certain reputation with the government.

> Having got clearance we could point to substantial clearance. And a reputation with the Ministry of Agriculture, Fisheries and Food [MAFF] that is substantial. And obviously that has helped our other food interests. We have shown ourselves to be a major, reliable, ethical food company, who are prepared to do with food the right things in the right way. Difficult to put a value on that, but I think it has a value.

The ICI toxicity and nutrition tests lasted eight years and it was important that the product ICI had shown to be safe was seen to be the same product that ICI produced. Additional toxicological tests were carried out over this period to 'track' changes to the product that came about through changed engineering requirements or for nutritional reasons and to show that these did not affect product safety.

BP had most of its testing performed in independent research institutes while ICI did its tests in-house. A manager commented on the BP strategy and how it was impossible to avoid all allegations that testing was biased in favour of passing the products as safe:

> BP . . . set up vast animal and toxicological testing facilities in the Netherlands. There was a small independent testing organisation, but by the time BP went in with its requirements it had grown massively. It was independent, despite other people's questioning of that. A 'no win' situation – if BP had done it themselves they'd have been criticised, equally when they funded and built up a big independent laboratory . . . [consisting] of all new and massive facilities . . . for multi-generation [tests on] chickens and pigs.

Private testing facilities were just not large enough to cope with the quantity of testing work that was necessary to show these products were safe. ICI and RHM had the alternative strategy of developing a good relationship with the regulatory authorities and performing toxicity tests in-house. The ICI toxicology testing manager compared the effects of this strategy to the Hoechst method of farming out testing to other research bodies:

> They had to share out work to government institutions and labs, Hoechst not having the right expertise, and so they were not able to discuss with the universities how their product might be tested. A specific example was a series of tests carried out by the Institute of Nutrition of the Veterinary School of Munich. They included Probion and Pruteen in fish diets up to 70 per cent as the sole protein source and showed a subsequent depression in the fish growth rate. This was published as a biological deficiency in the product, and it was the first publication on Probion. However, there were two or three previous ICI publications which already admitted that about 10 per cent addition to diet was OK for fish, but greater than 70 per cent led to depressed growth. This was important because it established ICI's reputation for veracity [with the authorities]. Also because the FDR would be receiving reports from their own institutions, not Hoechst, that were critical of the product – bad for the government–Hoechst relationship.

With an established test regime for novel food products, Dansk Bioprotein believe it will be easier to gain regulatory approval for their product. The Dansk Bioprotein head of R & D described his company's approach to the Danish regulatory authorities as one of 'negotiating for supplementary approval to Pruteen', In their documentation Dansk Bioprotein link the only completed toxicity tests on their product, by Steingass at the German University of Hohenheim, to ICI's work.

> ICI has produced bacterial protein basically from the same raw material . . . the results of Dr H. Steingass are in close agreement with results of an extensive experimental programme carried out by ICI in Germany and England and involving more than 300,000 animals. From this we conclude that Bioprotein and the ICI product are immediately comparable.
>
> (Dansk Bioprotein 1988b: 6)

Long-term feeding trials have been started with a Danish feedstuff company, but Dansk Bioprotein are making much less significant effort compared to ICI because they are not having to establish the testing regime itself. They have therefore cut out research that was a third of ICI's costs and if they succeed with their argument that their product is comparable to Pruteen, they may reduce the amount of testing that is required for their product still further.

The toxicity of the substrate and of nucleic acids

There were two issues in particular that were important in the tests. One was the toxicity of the substrate and the other was the effect of nucleic acids on animals or human beings. Most managers believed that a pure substrate was desirable for an SCP process, the only exception to this was an ex-BP consultant who defended the Lavera process based on gas-oil. There were no specific incidents or evidence that the Lavera process was affected by the toxicity of the substrate; nevertheless, two managers who worked on the BP alkane process expressed doubts about using such a substrate. As one expressed it, the Lavera product was safe, but using a mixed substrate 'wasn't the done thing'. A research manager at the IFP expressed these doubts:

> When you introduce a new product into the human food chain I believe it is always necessary to start from well-defined starting materials. And if there is a product that is not well defined, it is gas-oil. It is a mixture of products of different families and that poses great problems.

Ironically, the SCP product from Lavera was approved by the French government and consumed for three to four years by animals kept by French farmers, but the Italian plants, based on the purer paraffin substrates were publicly attacked because paraffins contain tiny proportions of carcinogens which remain after they have been refined from the truly mixed substrate of crude oil.

The nucleic acid issue was only a serious problem in Italy, where it was used to attack the safety of Toprina and Liquipron. There was a whole literature on human and animal response to different levels of nucleic acid intake, the problem being that the by-product (uric acid) of nucleic acid metabolism causes gout in people if it is produced in large enough quantities. Because micro-organisms grow so rapidly a high percentage of their cell mass is nucleic acid – DNA and RNA – and a diet rich in micro-organism-based products risks being high in nucleic acids and raising blood uric acid concentrations. The problem that the prospect of SCP as a human food raised was how much nucleic acid could be tolerated in human diets. A former BP consultant commented that 'The famous problem of nucleic acids . . . is a complete nonsense. Nucleic acids are already being metabolised by all animals. It may be a problem for direct human consumption'. If SCP was used as an animal feed, high nucleic acid levels could be tolerated because SCP would only be used mixed with other feeds. No extraction of nucleic acids was necessary. However, some toxicity tests were designed to examine the response of animals to high percentage SCP diets, and then toxic effects could occur. One of the Shell researchers thought that Shell had found something amiss with Pruteen:

> in the trials there was a major hiccup . . . there was a problem with Pruteen, it caused liver necrosis in chickens, due to the nucleic acids. We published this finding in an obscure journal to avoid embarrassing ICI. For legal reasons we published. You know you pick a little-read, second-rate journal, you give the article an obscure title, no one's ever going to find it. We talked to ICI about it, they had missed it somehow.

An ICI manager commented that 'there was worry about the [nucleic] acids in the company, but then we did a hell of a lot of toxicity trials to show there wasn't a problem'. One of the RHM managers commented that UNPAG eventually set an advisory maximum of nucleic acid content for human food products of 2 per cent by weight, a limit which was exceeded by all the novel food fermentation products. Hoechst solved this with the Schlingmann process for nucleic acid extraction, but RHM had a more difficult problem in that they needed to preserve the texture of their product. This problem of getting the nucleic acids out of the cells without destroying the product they compared to getting a car engine out through the grill in the front without opening the bonnet – nevertheless they succeeded and the mycoprotein product meets the UNPAG requirements.

NEW OPPORTUNITIES OPEN THROUGH THE DEVELOPMENT OF NOVEL FOODS TECHNOLOGY

As the companies built up their science and technology base in the pursuit of a commercial SCP process, other technical and commercial possibilities arose. Companies differed in whether and how they exploited these avenues, but whether or not they decided to develop such side issues, which were not strictly necessary to the development of SCP, there was always a legacy of fermentation-related skills and expertise.

Many managers thought that the SCP work had helped to develop the industrial skills and the projects that form today's biotechnology industry. One of those involved in the IFP projects thought this had been the case in France:

> I believe that the development of SCP has allowed the development of biotechnology in France. Because many of the teams that were created at that time have gone on working. Biotechnology is a small world, and one person in ten has worked on SCP. It allowed the development of this sector by training people. From this time the manufacturers were looking to develop electrodes resistant to sterilisation, pH electrodes, electrodes for monitoring oxygen concentrations, and so on, since oxygen was the main limiting factor in yield. A huge amount of engineering work, to create huge fermenters and sterile conditions . . . we learnt much about the scaling-up of fermenters.

One researcher who worked for the Kuwaiti government's Institute for Scientific Research believed the Kuwaitis gained from their experience of building sufficient expertise to be able to compare and judge the various SCP processes on offer from ICI, BP and Shell:

> They have not closed down what SCP really gave them. The research facilities still exist with the personnel and the expertise, they have a resource . . . They are now orientated towards making

biosurfactants for *in situ* cleaning of large oil tanks. They would never have got into biosurfactants if they had not had the physical and the manpower situation on the ground.

Although the skills that the companies built up were seen to benefit biotechnology in a general way, companies differed in the way they managed the skills that were left to them.

In Shell the microbiologists thought that a reorientation of research into the biotechnology area was a gain:

> I'm not sure, when you come to the crunch, what the experience gain was, more important was a feeling that . . . a knowledge that biotechnology was interesting. Not a bad word, quite a good word. SCP got us interested in biotechnology. It gave biotechnology a coat hanger . . . catalysts, enzymes, and so on. It could have gone the other way if there had been costs of £50 million they would have wanted to disassociate as rapidly as possible.

Paradoxically the early and inexpensive failure of SCP in Shell meant that the company would support researchers in their exploration of other biotechnologies, but in BP the costly Italian disaster led the board to dissociate itself from biotechnology as well as SCP. The idea that the BP board reacted against the idea of biotechnology was referred to in Chapter 5, but it took some years before BP disengaged totally. After the joint BP–ANIC company Italproteine had been liquidated most of the project's microbiology R & D personnel were transferred to BP's Sunbury research centre into a Biological Sciences group, which for a time held seven or eight researchers. Several researchers said that the biotechnology skills were then progressively lost.

> For a number of years we ran a much broader remit, with two or three colleagues at Grangemouth, but for the first five years a fermentation project remained. The remnants of the bacteria methanol process, we kept an interest in that. We were interested in a biopolymer, the same as Shell had done. Then we got interested for a couple of years in microbial products . . . and fuel-ethanol, because all round the world people were saying, make fuel from renewable resources. So BP decided, we need to suss this out. We said, you need to do some research, and two or three years of nice interesting work followed. Still fermentation, but different systems. But in the last four or five years that has also gone.

None of the fuel-ethanol research found projects likely to be economic – even though this kind of technology reverses the aim of SCP technology by producing fuels, from biological products as inputs. With the end of this work there was a dearth of possible projects that linked BP's core business to microbiology and by 1983 the position for biotechnology in the company looked bleak. One of the remaining microbiologists described BP at that time as a company which knew where it was going.

> BP Nutrition was doing nicely, with no need for research. There was no easy niche for biotechnology, which was what microbiologists were now called. There isn't a BP biotechnology process. What do they want with surfactants? Enhanced oil recovery? Well, we have to compete against chemical products. We looked at enhanced oil recovery, but it's not on for the kind of wells BP has. Only for watered out, shallow fields, where it doesn't matter if it goes wrong.

The Biological Sciences group was finally disbanded in 1987. Only one microbiologist was retained with a literature monitoring role, who felt that if they licensed the plant tomorrow to Russia,

> perhaps we could get three or four engineers back, if we paid them enough. My God, they'd have to pay me a lot. After ten years it's too difficult . . . they could have put it together again after two years.

After the liquidation of Italproteine BP had made the technical designs for the Sarroch plant into a collection which they still offer for sale at £100,000, but which, to date, has never been sold. The company's last microbiologist commented on this collection,

> The real innovations of the BP process were chemical engineering ones. The microbiology was relatively simple. But for those times there was the newest this, the biggest that. And a lot of that has still not been published, it still constitutes what BP considers it might still sell. The ten-volume tome. It's a pity, they would have been better to donate it to a sensible chemical engineering department five years ago. It would have helped biotechnology a bit . . . how the whole thing was set about, how you arrive at this absolutely amazing, huge plant.

The Pruteen project was seen by all ICI interviewees as the essential base from which emerged Biological Products, the existing base for biotechnological research, many of whose staff are former Pruteen researchers. Because ICI saw the technology as the core on which future products would be based they were vigorous in following up potential 'spin off' products from the Pruteen research. An example was the 'Deep Shaft' project which one manager described as, 'a Pruteen fermenter buried in the ground, using a multiple collection of organisms as a culture . . . which carries on a non-specific fermentation which has a high mass transfer'. This was the Pruteen fermenter used to dispose of effluent. Deep Shaft had been quite successful:

> about a hundred were licensed. Now we do a little research to provide information to the licensees. The rights have been sold to another company involved in the effluent disposal business. Better to leave it to an established waste treatment company than to put in all the management time involved in learning about a new business.

Another product that Biological Products is developing is a bio-degradable plastic called PHB, polyhydroxybutyrate. Certain bacteria, especially hydrogen-utilising species, can be fermented under conditions where they form granules of PHB in their cells as a means of storing energy. The beginning of this branch of research was attributed to one of the Pruteen researchers who had done a PhD on PHB and brought this interest into the company, where it remained when he left. The PHB work has now been 'spun off' into a development company called Marlborough Polymers.

Unlike most companies, ICI and Phillips Petroleum chose to experiment with the genetic engineering of their organisms. The Israeli SCP researcher Goldberg called this the largest-scale application of genetic engineering to date (Goldberg 1985), and an ICI manager described the work as follows:

> The programme was implemented about 1978 when the main plant had already been built and people were only just realising what the implications of genetic engineering were. . . . We did more research on that organism probably than any company has done on any organism. We understood it very well, its biochemistry, its physiology, and we found pretty early on that there were one or two things in the organism, which had they been the alternative, you would have seen an increase in the yield. . . . So what we did was to establish a facility at Leicester University, the ICI Leicester Lab. It's still going. Where the genetics of the organism would be worked on. We also established a facility at ICI Corporate Lab,

Runcorn, for genetics work on the organism. We also placed a contract outside with a genetic engineering company. In fact ICI succeeded in doing it . . . it was published in *Nature* in 1980, I think. This newly genetically engineered organism was produced. It never went into the commercial plant because it's one thing to produce it, another to test it.

ICI had come to understand the biochemical pathways of its organism so well that they could imagine how to improve these to enhance its performance in the fermenter. Work continued on this organism into the 1980s, so that there was a report of the successful transfer of an energy conserving gene from E Coli to the ICI bacteria (Bull, Holt and Lilly 1982). But the improvements in yield that could be attained were limited and nowhere near sufficient to reverse the changed economics of the process. According to one research manager, even if a more efficient bacterium had been used in the process, it would have increased dry mass yield by only 1 or 2 per cent.

One of the BP microbiologists described the BP conception of where the research was going as very clear and difficult to deflect down odd alleys. There had been an attempt to begin research into the genetic structure of the BP yeast by one of the project leaders, but this was stopped by higher management as a diversion.

Another company to benefit from the Pruteen project was John Brown Engineering. They had gone on to design, but not build, fermenters even larger than the Pruteen fermenter for other companies, something they could only do after the Pruteen experience. Their biotechnology skill base had been built up in all the key chemical engineering skills: project engineering, sterile equipment design, instruments and piping, and so on. A John Brown manager commented that

> Effectively what that means is, you can wrap up all the technology in legality . . . but you can't take away people's experience. We have well over a hundred people in this office with dozens of years of bioprocessing experience each, mainly starting with SCP, that have gone on to other things.

Within RHM, the mycoprotein project had been the vehicle by which sophisticated process control was brought into RHM.

> Most of it started there, it was the need to control this particular fermentation process, an order of magnitude tougher than any other process, which was the basis of all that learning. That learning and those people who are now applying it to other industries in RHM are guys who cut their teeth on the mycoprotein project.

Other techniques specific to the food industry were brought into RHM with the mycoprotein project:

> The point at which sensory evaluation, as a front-rank technique within RHM, and with it taste panelling, taste profiling, really started, it was this project, because we had to say, what will people think?

Uhde, the engineering construction company cooperating with Hoechst, suggested a Third World application of their SCP technology. This would be a non-sterile fermenter utilising a variety of natural carbon sources, such as sugar cane and molasses. The fermenter was built and consisted of a paddle wheel in a transportable tub, to which the substrate and inoculant would be added. Then the fermented mass would be decanted and fed to animals. Research scientists in Hoechst thought you could end up with a good quality yeast.

Uhde hoped that if this fermenter was run in three or four different places in the Third World they might get some government orders. This project was also funded by the West German government:

> The whole unit was sold to the German Ministry of Foreign Development. They wanted to use this plant as a political tool. Their responsibility, not ours. They wanted to give help to sugar cane production on the Upper Nile. And Sudan would have been the next step. You could move it from place to place as the harvesting is done and examine the conditions for fermentation in different places.

However, the project was not a success.

> It ended in the first place. It was mismanagement and a wrong decision to give it to Cairo University. Because an Egyptian professor will never touch an engine like that. Never. [They are] too academic. So it was a nice toy to look at, but it was never used. . . . No, really, it was just built into the sand, so the next sandstorm that came along, it was not working any more . . . then they needed an air pressure gauge for something on a different plant, so they took it out . . . it was a very simple tool of development aid, that was misused.

The West German government funded this project partly as a means of exploring fermenter technology and Hoechst used their fermenter to investigate fermentation as a means of producing enzymes, although the main purpose was always to investigate SCP production. While the explorative and innovative nature of the project was partly government intention, it was also linked to growth in the Hoechst biotechnology department in the late 1970s.

> At that time the department of biotechnology . . . was growing and we were thinking, we should not think in terms of only producing SCP, but to use this pilot plant for other purposes also. It could be used to break down the nucleic acids into their nucleotides, which could afterwards be used as spices. Another was lipids. What could we do with them? . . . We had a lot of different clever ideas coming from the chemists and microbiologists working on the unit.

These were similar ideas to those that BP and ICI had developed, independently and for different reasons; in one case as an attempt to diversify away from oil, in the other in an attempt to find viable products once Pruteen was uneconomic. Hoechst embarked on the same research path, of looking for alternative uses for the biomass, for another set of reasons. They, too, produced proteins which could be used as gelling, foaming and viscosity-increasing agents. They had a protein extender for addition to hamburgers and even began texturising the biomass through the development of a spinning process. They used the industrial contacts of two of their food chemists to approach Nestlé and other food companies to find out whether these proteins were of interest. Hoechst found, just as BP and ICI had before them, that there were always lower priced and established alternatives available.

One research manager believed there had been some real fermentation technique spin-off into Hoechst's antibiotic business, and that this was the enduring benefit to the company of its foray into SCP production.

The former head of fermentation research at SugarCo thought the outstanding spin-off from the SugarCo SCP project was the development of a new type of fermenter, the tower fermenter. It was

> SugarCo's money, my energy, that took it from being just a nutty idea to a genuine alternative method of fermenting. It is now in use commercially in the manufacture of things like vinegar, citric acid. And in Japan, many antibiotics. I don't claim personal credit for any of that, except that I was one of the catalysts, with Haydn-Davies and a guy called Gareth Morris. There are a series of key papers on the use of the tower fermenter for SCP, we got it to grow Aspergillus Niger as pellets. Basically it's a tube on end, you

pump your substrate up at the rate it is consumed. A steady state reactor. The mathematics is fearsome and I don't pretend to understand it, but we had a chemical engineer who did. That work led directly to a lot of those commercial processes, not by us, but by those other people. In the overall sense, not the SugarCo sense, it had a lot of economic benefit.

CONCLUSIONS

The socio-cognitive construction of technology

This chapter takes us beyond the debate about the synoptic model referred to in Chapter 1 where the function of the research department is to 'apply science' to a market need and so produce technology. In no company was innovation simply a matter of 'applying' basic science such as microbiology to an industrial problem to produce new technology. The content of the chapter reveals the social processes through which the technology was constructed.

A variety of social and technical concepts were assembled by managers to guide the design and construction of the technology. In the case of mixed cultures, aseptic fermentation and the tests required to 'prove' the safety of the products, it was what constituted safe practice that was at issue. The assembly of a set of selection criteria for organisms of yield, growth rate, productivity and cell density, was a process of matching organism characteristics with economic objectives and again there was an element of judgement at work influenced by social affiliation (to company, to skill background). The fermentation technology was a result of a combination of choices relating the circumstances of the company (control over substrate supply, understanding of core business), social (what constitutes a 'safe' product) and economic criteria to the existing technical knowledge base. Some of these selections of project criteria led the companies to break established patterns of behaviour and norms of achievement, such as the heuristic that only pure cultures were safe or that continuous fermentation was not possible.

The analysis of the innovation process shows the synoptic model discussed in Chapter 1 does not represent the complexity of the innovation process. The managers on each project were continuously creating, justifying and adapting their own rationales for what they were doing, and the cognitive process outlined in Chapter 1 is a better representation of what the managers were doing than the synoptic model. It makes sense to characterise the innovation process as both social and cognitive, and the intended outcome of this socio-cognitive process is new production technology.

In this socio-cognitive process, 'technical' knowledge is interwoven with 'social' criteria and knowledge of the firm and its environment. More important than these categories of knowledge ('social' or 'technical'), is the pattern of understanding (cause maps) managers develop and use to guide the creation of new technology; this inevitably includes both 'social' and 'technical' concepts.[6]

Physical artefacts embody human knowledge

It follows from the above that this new production technology with its associated operating practices embodies the knowledge assembled through the socio-cognitive innovation process. It is easy enough to accept that the cause maps are knowledge because they are possessed by people, but since they are used to give shape to new technology, this technology embodies this human knowledge. The relation between the market and the firm can be discussed in terms of this understanding of technology as knowledge, but first it should be clear that there are two dimensions to the embodiment of knowledge in new technology: the socially owned operating knowledge and the physically embodied knowledge that constitutes the production hardware.

The firm, production technology and the market

A simple definition of technology is 'knowledge related to some physical object', or more precisely, it is the socially conditioned knowledge of use of an object or tool – that is to say, technique. So even 'simple' objects like a table or chair require people to have the knowledge of how to use them, knowledge of use being a part of culture (aborigines would not necessarily know what to do with a chair before learning from the culture that produced such artefacts). If an object were of biological origin, this definition would suffice, but manufactured objects are the output of production technology. Production technology consists of the production hardware (e.g. machine tools) and, once again, associated knowledge of use, or what can be called production technique.

It is technique (socially conditioned knowledge of use of artefacts), and production technique (socially conditioned knowledge of use of production hardware), which is 'socially owned' in the form of individuals' cause maps. However, manufactured goods and the associated production hardware also represent knowledge, because they have been given structure by people with the purpose of relating these products to people's technique – that is, people's knowledge of how to use them. In this sense these artefacts can be thought to 'embody' knowledge, although it is not the 'actual' knowledge of use, since this is distributed amongst people outside the firm.

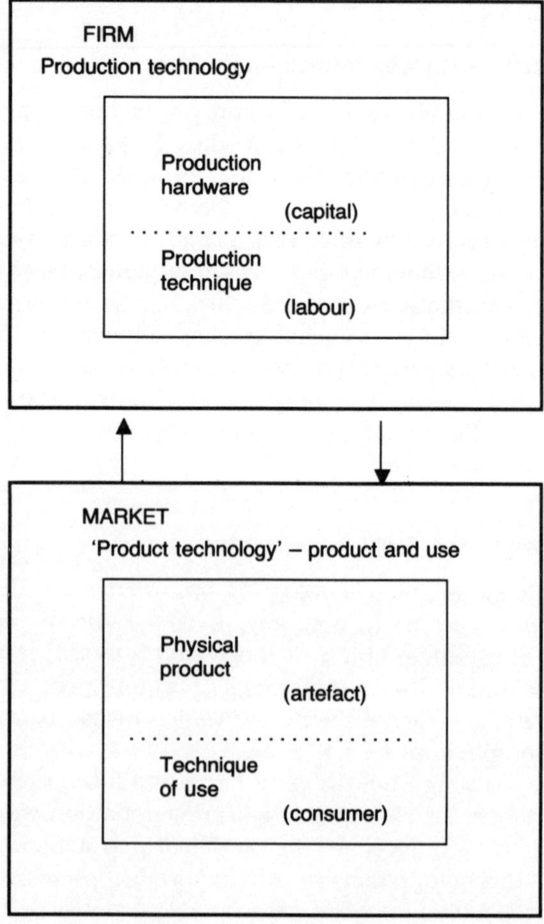

Figure 6.1 The relations between product and market, production technology and firm expressed within the key social division between the firm and the market. Innovation involves selective broaching of the firm–market division of knowledge

In Figure 6.1, the definition of technology in terms of knowledge is combined with the definition of the firm in terms of ownership of production technology. This diagram represents how production technology resides in the firm, while the knowledge of use of the product is possessed by those who make up the market. The individuals who make up the firm possess the cause maps which contain cognitions and rules of thumb relevant to the operation of the production technology which is the defining feature of the firm. The process of innovation involves relating anticipated technique of use to the sociocognitive construction of new production technology. This diagram emphasises that it is the social split between those who control production technology and those who use the artefacts and the difficulties this causes in the process of innovation that underlies the common characterisation of innovation as some kind of matching of technology to market need.

7 Project champions and project patrons

In interviews with ICI managers many references were made to the importance and role of a particular project champion on the Pruteen project who was also a member of the Agricultural Division board. The role of 'champions' was then pursued in interviews concerning other projects and the idea of the project 'patron' developed. This chapter considers what the term 'project champion' meant to the project managers and how the champion was seen to have affected project development.

One manager described Pruteen as perhaps an extreme case, where the champion was very closely personally identified with the project and its technology. He defined a project champion in this way:

> someone who works and fights for a project and takes every opportunity. He becomes identified with it. Most projects need championing or the inevitable difficulties will squash them into oblivion. They are needed from very early on.

There is an idea here that the commitment of individuals is necessary to override difficulties that less-motivated people would have allowed to end the project. This description also suggests that project difficulties do not have straightforward effects on a project; the significance of a 'difficulty' has to be interpreted by managers and that interpretation depends partly on their emotional commitment to the project. The idea that particular individuals can decisively affect a project's development through their personal interpretation of events recurred in many interviews: 'Who champions what is immensely important. Things happen because they are made to happen . . . not because they are right or wrong'. This comment suggests that although a project can be seen as 'right or wrong' (good or bad for the company), that is not sufficient for the project to take place. Development occurs when individuals commit themselves to a project, which may be because a project is self-evidently a good project for the company, or because in terms of that committed individual's interpretive framework the project is a 'good' project. This last possibility allows for others to see the project as a liability, and hence project politics can arise. A number of managers commented on what motivates people to make a personal commitment to a project:

> guys like Hart want to move into an area of technology that's growing . . . or they get frustrated. His internal motivation was, he wanted to leave a mark. I argued with him often, but respected him as an enormous driver and motivator, a leader of men. We were very targeted and very motivated, we had a great project champion . . . he motivated us.

The personal ambitions of the project champion referred to here involved doing something out of the ordinary, beyond the usual requirements of the company. He was able to communicate his belief in the value of the Pruteen project to many of those around him, and because he was seen as the origin of the belief that this was a good project, because he was personally identified with the advancement of the Pruteen project, he was called the project champion.

In the other projects it became clear that the project champion was not necessarily someone who was easy to identify or consistently important at all stages of a project, nor need there be only one champion for a project. Rather, the larger the number of interventions made on the project's behalf, the greater the identification of the person with the case for the project's success, and the more deserved was the title of champion. Project champions 'emerged' the more frequently they were seen to have contributed to the project's continuation – but not necessarily its commercial success.

Interventions included the origination of the project idea, intervention on behalf of the project when it might otherwise have been closed down and the support of the project through the subversion of formal company control mechanisms.

ORIGINATORS

The former chairman of RHM, Lord Rank, initiated and then protected the mycoprotein project for economic and philanthropic reasons. According to one of the current managers on this project,

In truth, the real reason RHM got into this area was because its chairman Lord Rank essentially combined two views of the world. The first view was attached to the fact that he had something of a philanthropic view of life. He and the Rank family had made significant investments in various Methodist charities at home and abroad. One of the areas that made an impact on him was what he believed to be an impending world food shortage, which he interpreted as a protein shortage. . . . He had an emotional commitment to doing something about that. The other thing was that he was a very, very hard-nosed commercial individual who appreciated that he had a series of flour mill investments . . . the likelihood was that they would be producing surplus starch. He was looking for a use for this starch.

The personal motivation of Lord Rank and his non-commercial interests are clearly important in the selection of this project. Lord Rank used his authority to start the mycoprotein R & D project, then to favour and protect it, but he was not involved in its day-to-day management. This role might better be described as that of project patron. A research scientist working on the mycoprotein project commented on the effect of the patron's support:

It was Lord Rank's pet. It wasn't somebody at the bench writing a project proposal and saying, have a look at that, it was actually the chairman, his hobby horse, and once it evolved, once it got past a certain time, it probably took a great deal of guts to say, no more, back!

Lord Rank approached the research department with his idea and the head of the department, Arnold Spicer, took it up with enthusiasm, so that one of the current managers called him the co-champion. Lord Rank and Spicer then protected the mycoprotein project in its early stages, and when they retired the same protective role was taken up by RHM's new chairman and new research director.

The BP research scientist Champagnat is referred to by a number of press and journal articles as the originator of the BP yeast-hydrocarbon processes (e.g. Laine and Snell 1976) and his name was often referred to by BP managers recalling the start of the Toprina project. There was another scientist who acted as a consultant to BP in the early stages of the project who disputed the BP version of the project's origin. He was a senior microbiology researcher in a French government research institute in Marseilles and suggested that it was BP's desire to minimise his own role in the early stages of the research that led BP to present Champagnat as sole originator. BP would not have done his role justice because he was not a BP scientist.

> From the very beginning I never patented my ideas, because if I had done so, I was afraid they would put it in a drawer and never go ahead with it. And they were afraid I would do so. One day an important person told me, we don't want a new Mr Gulbenkian, a Mr 5 per cent – who claimed to have 5 per cent on all oil produced in the Middle East. Then one day Champagnat wrote a book saying the idea came from him . . . he had never heard about microbiology before.

The accepted BP version of the story has Champagnat as inventor of the idea when he was working on ways of dewaxing Libyan crude oil. This was the view of all those interviewed except for this government scientist, who believed that Champagnat took the idea of deliberately growing SCP from himself during a long luncheon after the successful conclusion of some collaborative work on biological waste disposal.

The conflicting accounts of responsibility for the origin of the SCP idea are similar to academic scientific disputes over priority of discovery and derive from much the same motives of personal ambition and the desire to be associated with a success. This senior French government researcher identified himself as the 'father of SCP' and understood that BP had reasons to minimise his role.

> It was always important for BP not to depend too much on me . . . I have no resentment against that. I did not belong to the staff of BP, I was independent. You can understand that they wanted a separation.

Whether or not one believes this government research scientist, this dispute shows that where the project is seen as successful, the prestige of being identified as project champion is desirable.

THE SUPPORT FROM PROJECT CHAMPIONS

Some managers identified times when projects would have been closed if not for the protection of a patron or a champion.

> These two guys shared the vision, that without being able to say exactly what the market would be, without being able to say exactly what the economics would be and why the hell they should be doing it, they had an innate feeling, or belief, that this project

was different, was worth continuing. And they just protected it from all comers for 15 years. Nobody got near enough to cancel it. Had we asked the Board [of RHM], 'What do you guys think, should we continue with this?', they'd all have said no. Boards are the great bastions of conservatism, as you know. These two guys said, well, we just won't tell them.

So powerful individuals are able, to an extent, to impose their judgements in spite of company rules and the formal decision-making structures. This manager went further in stressing the importance of personal commitment to project development, using Pilkington's development of the float glass process as an example where the champion, Pilkington, was powerful enough to protect the research against internal criticism until it proved successful: 'You go anywhere near the other big decisions and you'll find somebody who put themselves in front of it and said, "This is mine, you are not going anywhere near it"'. The SCP project on carob bean husk substrate was kept alive by the head of research in SugarCo. The head of fermentation recalled that

> I inherited an SCP project. I entered a situation with a new job where it was very clear that I could not stop it. So I tried to modify it. Certainly, scientifically, that was a successful strategy. Economically and politically the numbers don't change. No market for the thing from the beginning.

The project eventually faded away, but,

> we could have decided that earlier, had Finlay not been such a champion. Almost as if you have to accept, when you run the personality cult, which is what we are talking about, that the man at the top, precisely because he is so dominant, precisely because he can get the funding, can be your champion, at the same time, will have his own pet projects, they may be non-runners. You just have to count that as part of the cost for the winners.

So the reverse side of the champion who fights for successful projects is the champion who retains hopeless cases, protecting these with his authority in exactly the same way as he can protect those that turn out to be successes. Another SugarCo manager commented that

> You see I think textbook analyses are often too stereotyped. Things don't happen in companies according to strict rules . . . and we decide we're going to do this and therefore that happens . . . it doesn't! It's very often the players and the game – the

players in the company decide these things and then the players change, personalities change, and companies really reflect the people who are running them.

The Pruteen project champion was important in the process by which Agricultural Division won approval for construction of a full-scale plant from the main ICI board.

The critical days were in 1974 and 1975 when the capital expenditure proposals were made and there was a continual visitation of main-board directors to Billingham. They came in their ones, twos, at monthly interviews, you are continually giving them the story, the background behind it, and it was orchestrated, Hart was the first butt of any questions, it was he who had to evaluate the pros and cons. So by 1976 when the expenditure proposal went forward he presented the case which represented the majority of the views of the people who were working on the project. He represented those views to the board, all the presentations from 1974 onwards.

Hart was crucial to the development of Pruteen by his real interest in all aspects of the project. One research manager commented on how Hart took a close interest in the research for which he was responsible:

Hart was interested in our research. In 24 hours he would know the result of our meetings even if he wasn't present himself. He would get the staff to sum up results in a five-minute conversation.

Hart's close interest partly explains the high motivation of managers working on Pruteen. He did not have the authority to take an investment decision himself, but his enthusiasm and commitment to the project were thought by many to have helped persuade main-board members to support Pruteen.

A senior Shell manager, Rothschild, had introduced biological science research into Shell, and so made SCP research possible:

he of course came from a biological background and had a lot of contacts in the biological fields. His idea was certainly to upgrade the level of basic scientific research in Shell.

This he did by building up two research centres:

the one down at Sittingbourne was very much oriented towards enzymology, biotechnology, microbiology, biochemistry, and yes, he saw obvious opportunities for developing the biological sciences in Shell.

Work on the methane fermentation project was carried out at Sittingbourne. Rothschild was described by one project worker as taking a close personal interest in the SCP project and encouraging it. But he played no part in the research decisions themselves, nor did he make a personal commitment to its success – it was a promising project to be judged within the usual formal procedures. When asked if Rothschild was a champion, one researcher agreed enthusiastically, while another had doubts about who, if anyone, had been a champion of this project, while acknowledging a need for them:

> they will only allow you to nurture a project for so long before they begin to ask questions . . . we had to find a home for it . . . a lot depended on having champions. I suppose Rothschild was one, but you need to have one at a lower level. Probably no one after Rothschild. Even if Rothschild had been here he would have asked questions. He was in no position to authorise investment in any case.

So Rothschild was critical in creating an enabling environment that allowed SCP research at Shell, but he wasn't seen to commit himself personally to this project's development, so perhaps he should be seen as a 'partial project patron'. One of the SCP researchers at Shell was personally and closely associated with the project, but he did not have the authority to advance it and,

> I left because I had fought so hard for the project, I was known as 'Mr SCP'. If SCP did not exist, it was rather hard to maintain that existence. I left because there were no career opportunities within the company. They treated me, always extremely well, I could never complain about salary, but where do you go?

Although this researcher could be seen as a champion, he did not have the seniority to advance the project and when it closed he had a problem because he was then associated with a failed project and he was unable to convince the company that he was more than just an SCP specialist. In order to progress he felt he had to leave the company, which itself was not an easy option because of what he described as the attitude of academic institutions to Shell.

> Whenever I applied for a job in the academic sector people said, you can't possibly want to come here, the resources you have are second to none in the world. All you are doing coming here is to get us to give you an offer so you can bid up your salary. They [the academic institutions] wouldn't even make an offer.

It was sometimes difficult to escape the taint of SCP. The project champion in ICI was also widely perceived to have suffered from the failure of the project with which he had come to be identified and he left ICI in the early 1980s. If Pruteen had succeeded he might still be within the company. One manager from outside ICI interpreted this early retirement as follows:

> Hart carried Pruteen through and when it failed, he got it in the neck. If you speak to him now, he'll say, it [Pruteen project] was the right decision. Because without it, ICI would not be in Biological Products.

The identification of a personality with a project ties the career of that individual to the project very closely. The SugarCo carob project gives an example of the converse: a project dying away when the sponsor moves on to new tasks. When appointed, the head of fermentation felt that he had no choice but to run the carob project he had inherited from his head of department, but he modified it so that it was a scientifically interesting project for himself and his staff, but

> What really happened to SCP was that long before SugarCo said we are closing down all this it died the death anyway. The champion in the shape of myself was busy doing something else.

He was sure that the carob project would have gone in the great squeeze on research after 1979, but it happened several years before because

> the man who took over from me with responsibility for the fermentation project was a guy I had recruited and he was a different kind of microbiologist from me. He had come from Glaxo . . . he was left with SCP. He de-emphasised it rapidly. He had to finish his commitment, he did a fine job on the Belize project, writing it all up . . . I was not going to be a scientist any more so he imposed his own research thinking. He was picking up the leftovers, and the last thing he wanted was SCP.

According to the head of fermentation, the major run-down of research in SugarCo in 1979 prompted the champion of the research diversification policy, to leave the company:

> he left a year later, a consequence of this surgery, since he had been the champion, and since people like me had already gone, I think he took the view that he would take honourable retirement.

THE SOCIAL CONTEXT OF CHAMPIONS

Managers modified their descriptions of the prominence of the champions they were describing by accounts of the network of advisers they used, part informal and part formal. Hart took an unusually close interest in the research on Pruteen, but he also had a more formal network of advisers:

> There was a team advising him, he had a team of people on all aspects of the project. They met regularly to discuss it. He took a very, very close, day-to-day interest in the project. So he was not making decisions in isolation, but on the basis of what he was told by research, business area, economics, engineers, everyone.

The quality of the people around the champion was picked out as critical by the present-day management of the Cellulose Attisholz paper mill, when they talked of the success of the mill's founder, Sieber:

> you need someone with the foresight, who is influenced by others from R & D and marketing. But the man at the top has to choose the right people to have around him. Needs a vision.

The quality of the group advising the champion is important and few of the decisions of the champion will not be influenced by those around, even if these advisers have less prominent names and lower profiles. The importance of the champion's social context tends to erode the idea of the champion as a 'heroic' leader forging decisions from their own resources. The pattern of social interaction around and with the champion, which may have been designed by the champion for the current project, is the mechanism by which the champion develops and guides their decisions.

CONCLUDING COMMENTS

Project champions are managers who become personally identified with a project, generally because they exceed the expected level of commitment and seek to persuade and argue the value of the project to the company. Patrons are likewise identified with a project because they use their authority and power in the firm to encourage its development and to protect it.

Project champions are important not only at the 'giant project' level, as with Pruteen. The case of the carob bean project suggests that there is often a degree of personal support required for a project to continue on the experimental, low-cost level and that this will vary depending on who is in charge of a group of projects or a department.

Champions are probably better thought of as 'emergent', since the perception that someone is a champion develops as those around identify a pattern of action that consistently supports a project. The degree of identification increases the more frequently they use their influence to advance the interests of the project compared to those around them. So there can be intermediate levels of commitment and 'partial' champions.

One manager referred to a personality type who would look for a project to champion – this is the view that there are personal characteristics, such as ambition, which make some individuals more likely to become champions than others. However, it was clear that there was a situation where the commitment of the champion was linked to the degree to which the project served their core beliefs, or cause map. These are not two opposed views – ambition varies between individuals and so do the cause maps which create successful projects. It is in the champion's cause map that we can expect to find the reasons for the champion's selection of a particular project.

The champion is committed to a project because of the high degree of match between their cause map of understanding of what is a good project, what is good for the company and for themselves. Other managers have different cause maps, and different selection rules for what makes good projects, and are more likely to take the role of neutrals or sceptics, if the organisational structure allows them to do so.

In the case of Pruteen the following three, stylised beliefs were identified by managers as factors that predisposed the division director to choose to champion the Pruteen project. These were:

1 Methanol would become an alternative world chemical feedstock to ethene with a consequent long-term drop in price.
2 A protein crisis would develop in the long term as protein demand outstripped supply.
3 Fermentation technology was a key part of the developing cluster of biotechnologies and the development of an early lead could lead to huge business opportunities for ICI in the future.

As elements of understanding these were not unique to Hart, but in combination they begin to define his particular cause map. The Pruteen project met these conditions and was therefore likely to attract the support of those with these beliefs.

Many other cognitive elements exist which affect the champion's thinking, but most will be common to many other managers and serve rather to identify those working on the project or those who are members of the firm, for example a background training in micro-biology, an acceptance of commonly available market size estimates for the new product, an understanding of the soya market and a knowledge of how a project is judged by senior management. However, it is the unique combination of these common elements of understanding with some of the rarer elements that creates the particular cognitive understanding of a champion and which allows certain individuals to see in particular projects a way of advancing both themselves and their world.

This way of conceiving of champions does undermine the idea that champions are solely personality types and the related idea that they could be selected by psychological testing and then given a project 'to champion'. The phenomenon of the champion is an extreme version of the natural tendency for individuals to prefer projects which strongly match their core beliefs. This account of the champion idea is also compatible with Elger's picture of the organisational world as based on a micropolitical process; champions are those who have successfully used this process to convince others that their way of seeing the world is right, or better than the alternatives (Elger 1975; see also Chapter 1 of the present volume). This view also gives us an insight into the difference between champion success and project success. A successful champion will have persuaded his company to develop his project, but that does not imply the project will be commercially successful. It only tells us something about the individual and the micropolitical process in the company. The project may go on to fail commercially and by doing so affect the champion's position in the company, although the appropriation for blame will also be an outcome of the micropolitical process.

Freeman has referred to the role of advocacy and persuasion in decision-making on projects:

> Empirical evidence confirms that decision-making in relation to R & D projects or general strategy is usually a matter of controversy within the firm. The general uncertainty means that many different views may be held and the situation is typically one of advocacy and political debate in which project estimates are used by interest groups to buttress a particular point of view. Evaluation and technological forecasting, like tribal war dances, play a very important part in energising and organising.

> (Freeman 1982: 167)

The use of such formal cost-control methods is entirely consistent with a micropolitical process. The methods themselves do not resolve political problems but may contribute to the structure of the debate that does lead to a consensus behind the case for a particular project. In the micropolitical process, support may still be given because of the ascendancy of a champion or a coalition of interest groups.

Weick also refers to this micropolitical process when he suggests how negotiation within the firm tends to homogenise the firm's view of the world, that is, establish a common interpretation of events, a common cause map (Weick 1979). Part of the evidence above is that this process is never completed, nor is it likely to be. Some managers might disagree with the champion's cause interpretation of events, but they might not benefit personally from vocalising dissent, while the champion does not need to negotiate a common understanding throughout the company, only with those individuals who have the power to confer sufficient resources to continue the development of a project. A company-wide cause map is not necessary for one cause map to come to dominate one of the company's activities.

Projects do not necessarily need champions, as was shown in the BP case. When the highest levels of the company collectively back a project from its earliest stages, individuals at lower management levels tend not to become prominent. To a degree the champion exists because there is the opportunity for persuading sceptics, especially sceptical senior management of the value of a way of thinking about a project. If the champion is convincing, the project wins company resources, but the final test of this extension of trust to the champion is the market place, where with product success or failure the champion's individual position may be reinforced, or damaged to the extent that they leave the company. Within BP the decision to develop Toprina was a board decision and required no lower-level individual to argue and push for the higher levels to support the project. There were senior executive managers who made their careers on the project, but they were not advancing the project through a personal commitment – they were doing as the board wanted.

CHAMPIONS AND PATRONS IN THE LITERATURE

The understanding of the project champion outlined above and summarised in Box 7.1 can be compared with the 'heroic entrepreneur' model of innovation development, where 'great men' have the unique personal qualities of the entrepreneur which enable them to forge new companies and industries – for example, in Schumpeter's first ideas on innovation and long waves (see Chapter 1). In this model a number of features of the role of the individual in innovation are ignored. These may be summarised as follows:

1 The personal motivation of the entrepreneur may be more complex than a simple desire to make huge profits.
2 Why certain projects are selected over others is not explained – it is assumed that their value is evident before the case for their value is made.
3 The role of champions within corporations, that is entrepreneurs within a complex social context, tends to be ignored.

As far as the firm is concerned, this cognitive explanation of the champion implies that effort spent identifying the personality characteristics that make a good champion isolated from social context may be misplaced. In any case, champions do not guarantee project success – the firm might better investigate ways of exploring and assessing the belief systems of champions than simply cultivating the champion phenomenon.

Box 7.1 Summary points on project patrons and champions

Patrons and champions

- Patrons may use their power in a company to advance a project; they take 'personal interest' in a project.
- Champions commit their time and careers to advance a project by persuading others in the company that it is worth supporting.
- Champions and patrons are identified with a project because they are perceived to work for it more actively than the norm.
- They are seen to be personally responsible for advancing a project and gain or lose personally with the project's commercial success or failure.

Reasons for a champion's personal commitment

- Personality – they may have personal ambitions or reasons for their commitment to the project.
- Their personal interpretation of the project's nature and value to the company (their cause map) may differ from their colleagues' interpretations, leading to a 'belief' in the project.

Champions are emergent

- They do not necessarily exist as 'personality types' before being linked to a particular project, i.e. they may emerge as champions as a project continues.

- They need not be dominant throughout the life of a project.
- There could be several outstanding and identified personal contributions made to a project in its lifetime, or one, or none.
- Selection and championing of projects is part of the continuing micropolitical process inside the firm.

Champions and the firm

- A wide range of potential projects can only be conceived, or brought into the firm for discussion by individuals with diverse ways of seeing the world, i.e. a diversity of cause maps.
- Championing of a project by the board, a subgroup within the firm or an individual is a necessary condition for the project to be developed by the firm, whether or not it subsequently proves successful.
- Cause maps modified through the experience of success and failure in projects lead to a change in the prospective future selection of projects and champions.

The evidence in the chapter can be used to make the simple point that the presence of a champion may not necessarily lead to commercial success, although we might conclude that the chances of successful development of a project are increased if some part of the firm with authority (board, patron or champion) believes in a project so that they will support it against short-term set-backs.

There is some additional evidence that the presence of champions correlates with commercial success. For example, one of the results of Project SAPPHO (Science Policy Research Unit 1972) was that commercially successful projects were linked to the activity of a champion more often than commercially unsuccessful innovations.

In summary, then, project champions are the corporate equivalent to the company-founding entrepreneur; they are the manifestation of individual initiative within the company and in this chapter we have shown that they are identified as such within the company by their belief structures as much as by any personal traits they may have. Whatever their personal motivations, the way they appropriate company resource is through a micropolitical process of negotiating a common understanding with sufficient individuals in authority within the firm to enable project development to proceed.

8 A socio-cognitive approach to innovation

After a review of earlier findings this chapter explores some of the connections between these and the ideas on technical innovation outlined in Chapter 1. It proposes that the understanding of innovation as a socio-cognitive process, developed through analysis at a micro-level, is compatible with the macro-level features of technical change.

REVIEW OF CONCLUSIONS

The market concept

In the innovation process the market is a concept, a mental construct, whose construction can be changed by those involved in the innovative project; but to an extent the development of this construct is directed by the establishment of links with potential users outside the firm.

As a result of examining how the market for a novel product can be conceived of before it is traded, it was found useful to distinguish between the market concept for the developing innovation, the 'innovation market concept' and an associated market concept for a product already traded called the 'reference market'. The innovation market concept included elements of understanding[1] that the reference market did not (some contributed by the R & D department).

The process of construction of the market concept was therefore both social and cognitive: social since selection of relevant social groups external to the firm and the use of the R & D organisation provided the cognitive resources for the development of the innovation market concept; cognitive because there was a process of selection of knowledge from these groups.

Competition as a socially negotiated concept

In Chapter 4 the lack of a clear mechanism by which the idea of 'competition' could influence the projects was striking. There was no straightforward 'cut-throat' market share competition in SCP development, rather competition occurred at the same time as cooperation did on other issues. Managers were having to negotiate an understanding of which of their activities should be secret, which could benefit from cooperation. In the innovation process what is thought to be of competitive significance is a matter for negotiation and interpretation.

Firm culture and senior management

This chapter examined how structures both formal and informal and internal to the firm were perceived to influence the innovation process. These included the role, status and operation of the R & D department, the perceived bias in interpretation that resulted from certain skill backgrounds, the influence of the boards and senior management on the projects and the role of project champions (see Chapter 7). Common to all these structures was that managers perceived them to influence the process of innovation.

Technology as a form of socially structured knowledge

What were at first sight technical choices necessary for the projects were examined and found to be closely related to commercial and social considerations. Structure was given to technology by drawing on firm culture, strategy, conception of the market, the technical base of the firm and even influenced by the background and experience of certain individuals. This process was called the social construction of technology and its output was new production technology; new hardware and associated operating technique. The new production technology embodies knowledge that was assembled through the socio-cognitive innovation process.

A SOCIO-COGNITIVE APPROACH TO INNOVATION

We can use Weick's idea that people use their own theories to interpret and understand the world (see Chapter 1) and Turner's suggestion that theory generation is a process of assembling cognitive elements into a coherent framework (Turner 1981) to describe the process of innovation. Those involved in innovation continuously select cognitive elements from the market and technical base of their firm to create what they believe to be a coherent project conception, which guides the social construction of new production technology intended to meet perceived market need. They work within a framework of constraining and enabling structures such as firm culture, structure of the firm, perceptions of the competitive environment, which influence the project where they are included as cognitive elements in the cause maps of project managers. The conception of the entire project can be portrayed as the bringing together of the conceptions of innovation market, technical base of the firm, firm culture and structure and we can refer to the result as a 'cognitive ensemble' because each of the constituent areas are cognitive constructs. The term is useful because the 'project idea' is judged both as to how its constituent assumptions match existing belief and practice and by how far it forms a coherent group of ideas.

Social process is an integral part of managing innovation. At its most basic it includes the negotiation process referred to in Chapter 1, where common cognitive understandings of the project are developed between individuals (although, as emphasised in Chapter 1, it may only be a partial common understanding, even on the project itself). Social process arises again where the innovating firm attempts to tap cognitive resources socially located elsewhere; for example, in the effort to develop the innovation market concept relationships were formed with selected other firms (e.g. feed compounders). Internal organisational structures such as the R & D department can also be seen as providing original cognitive resources to aid the process of matching innovation market concept to new production technology design.

This analysis can be described as a socio-cognitive approach to the innovation process and it can be combined with other concepts outlined in Chapters 1 and 6 – for example, Teece's view that the firm is defined by its possession of specific production technology.

A SOCIO-COGNITIVE VIEW OF THE FIRM

An essential feature of what we can call a socio-cognitive model of technical innovation in the firm is the split between the social group which manages production technology and the social group which possesses technique of use of artefacts (see Chapter 6). The split itself can be seen as economical in terms of individuals' bounded rationality since technique of use is less complex than production technique and requires less synergy of interaction (less complex organisation). The individual as user may then have a range of techniques of use for many artefacts while only possessing partial knowledge of production technique in their role as member of a firm (hence the need for 'ease of use' in a product).

An innovatory project begins as something which is a series of linked mental cognitions, or 'cognitive ensemble', which is innovatory in its combination of selected cognitive elements. The creative process is not required to generate novelty from 'nothing'; novelty becomes a relative and subjective judgement of the ensemble and innovation is linked to existing technology, products and markets through the cognitive ensemble. As people and resources are committed to the development of the idea, the set of cognitions that formed the original idea changes. The personnel engaged in managing an innovatory project are involved in selecting the appropriate cognitions and rules of thumb to define the project in relation to the firm and its technical base. The project definition has a degree of coherence between individuals, because they negotiate a consensus on such definitions. However, it follows from the social separation of production technology from artefact users that the most important characterisation of project development is the 'dialogue' between the technological base of the firm and the market concept. The idea of a dialogue is useful because, as the production technology and the market conception acquire detail, each may require change in the conception of the other.

When the firm embarks on an innovatory project, it is with a developing understanding of the likely changes in the market as a consequence of its innovation, and this involves understanding the changes in how the product may be used. At the same time, the firm must adapt its technical base to these perceived needs of the market. It is an iterative process of definition of the market–technology relationship – managers using a market definition to select the development of new technical routes, and then refining how the market will be affected by their choice of technical development.

The idea of a market is just that; the 'market' is a conception – a mental construction composed of cognitive elements brought together to form a coherent market conception for the projected product. It is at once a model used to guide the construction of the production technology, ensuring that desired user technique is built into the product; but it also gives the innovating firm confidence that its product will succeed, since it enables comparison of the innovatory product with existing products in terms of the cognitive elements which make up the market concept. It is also the case that it is not possible *not* to have a market conception during innovation, in the sense of an imagined use for the projected product; how rigorously that conception is developed and matched with developing technology through surveys, interviews or talk with users is a matter of judgement for those within the firm. The developing production technology is intended to embody the developing market concept.

While the market–technology relationship may be central to the innovation process, these terms are broad enough so that other characteristics of the firm are included in the moulding of the final choice of production technology and market served; these range from strategy, culture, the role of individuals, firm size and resource for innovation and the like. These characteristics enter the cognitive ensemble, just as do the market and technical cognitive elements.

In Chapter 1 there was some discussion of Teece's idea that the firm could be characterised by its technical knowledge base; here we have seen the process by which the knowledge base is changed. Creation of new SCP production technology altered the skill base (through the importation of microbiological skills and their synthesis with engineering and other skills) and the hardware possessed by the firm and so incrementally shifted the knowledge base. Other biotechnology projects were possible, and indeed planned, and these would have further shifted the knowledge base and given rise to a biotechnology 'trajectory' of the sort described by Teece. Other of Teece's characterisations of technical knowledge were evident; for example, knowledge was transferred through transferring people, and when the skilled people left a company (BP) the company was perceived to have lost the ability to innovate in this area (despite the existence of a formal, written 'tome' detailing how to build the plant).

This is the socio-cognitive process of technical innovation in the firm: it combines the ideas of individuals thinking cognitively, technology as knowledge and the firm defined through its possession of production technology.

FROM MICRO TO MACRO

This process at the micro-level should be compatible with macro-level features of technical innovation, such as the long waves described in Chapter 1. These are periods of economic restructuring which Freeman and Perez (1988) associate with clusters of technical and organisational innovation. In the case of SCP, had hydrocarbon prices remained stable and low, there is little doubt that substantial (billions of pounds) of investment would have been made in the technology. If this kind of 'clustering' is imagined to take place on an even larger scale, then the aggregate investment pattern of a long wave might result.

The pattern of long waves at the macro-level would arise if there was a patterning of the socio-cognitive innovation process across many firms in the economy. If there were common cognitive inputs to many firms, then their apparently independent investment decisions would be similar because of this common cognitive basis. Variation in one of the common cognitive elements (such as a lowering of hydrocarbon prices) would serve to trigger investment across many firms, so aggregate investment would tend to be 'bunched' in time. At the macro-level one might be content to think that the lower price level had 'caused' the investment, but from our perspective it should be clear that the investment only occurs as far as there are firms and managers in a position to interpret and use the fact of the lower hydrocarbon price in terms of the innovatory project concept (cognitive ensemble).

Another way in which the irregular 'bunching' of investment that underlies the long-wave pattern may occur is through the creation of new technology itself, which makes available to other firms a cognitive resource. This may trigger investment in a further series of innovatory projects as the new technology becomes incorporated into these other firms' cognitive ensembles. It is illustrated in the example of Dansk Bioprotein, whose project was only possible at all because of the innovations made by the equipment manufacturers since the 1970s, for example in centrifuge and spray-drier technology (see Chapter 6). The socio-cognitive nature of the innovation process is essential to the operation of this 'feedback' system.

In general, one wave's ensembles will become embodied in a range of products and processes which, as they diffuse through the economy, change the cognitive environment from which are drawn the elements of new cognitive ensembles.

Appendix

ECONOMICS

Dansk Bioprotein (natural gas)

The Danish venture capital company Dansk Bioprotein claims that SCP for animal feed is once again economic and the company have proposed to build a semi-industrial plant to produce SCP for the veal calf milk replacer market and perhaps even the piglet feed market (see Chapter 3). Dansk Bioprotein have published figures shown in Table A.1. So it appears that as a milk powder replacer in animal feed, Bioprotein is an economic proposition, irrespective of the price of fish-meal (fish-meal cannot be used to feed veal calves). Dansk is fortunate here, because the EC removed much of the subsidy for the use of skimmed-milk powder as animal feed in 1987, so recreating the market for SCP as a calf feed. If fish-meal prices remain high and we accept the nutritional equivalence of Bioprotein and fish-meal, then the large animal feed markets are viable propositions for Bioprotein. Dansk estimate that the value of the replacement market for Danish piglet feed alone is 68,000 t/a and their proposed plant is to have a capacity of 18,250 t/a, so its output could be taken by the piglet feed market alone.

Given the major changes in economic variables, the use of a different technology in the Dansk process is itself probably not the

Table A.1 Cost of Bioprotein compared to cost of dried milk powder

Cost of Bioprotein (DKr)	Cost of dried milk powder (DKr)	Cost of fish-meal (DKr)
4.9	12	4.12 May 1988
(£417/tonne)		5.78 Oct. 1988

Source: Dansk Bioprotein 1988b: 14

major reason for Dansk Bioprotein's viability *vis à vis* Pruteen and other methanol-based SCP ventures, but if the Dansk Bioprotein plant is economic, then the Pruteen plant certainly would be. The prospects for the plant are outlined in Chapter 9.

GAS-OIL PROCESS

The 1986 decrease in the oil price has improved the economics of SCP processes based on hydrocarbons such as gas-oil; prices have been halved from near $30/barrel to near $15/barrel and in real terms they are at a post-1973 low. One of the BP consultants believed that the BP Lavera gas-oil process would be economic once more. A crude estimate of costs suggests that this is so.

Gas-oil   £94/tonne
Soya meal   £150/tonne
(source: *Financial Times* commodities index)

Fish-meal   £400/tonne approx
(source: Dansk Bioprotein 1988, May 1988 price and with DKr
             12 to £1)

If we assume the gas-oil price represents 50 per cent of operating costs per tonne of SCP and that the yield of SCP from one tonne of gas-oil is one tonne, then the cost of producing one tonne of SCP is $1.5 \times £94 = £141$.

Whether the plant is viable or not depends on the criteria chosen for a return on the capital investment; for example with a 100,000t/a plant cost of approximately £100 million (1989 prices) and a severe three-year payback period, £33 million return in each of three years will be required, adding £330 to the price of each tonne produced and giving an uncompetitive SCP price of

$$£330 + £141 = £471/tonne$$

More lenient capital return criteria make the investment look attractive as a simple fodder yeast. If sold as a veal calf milk replacer, as BP France sold it in the early 1970s, it looks profitable. The Lavera process was reportedly still able to cover costs in this market after the 1973 price rise, so it is not too surprising that it should once more be economic.

These crude calculations suggest that both a gas-oil and a natural-gas-based process would be viable under present economic conditions, if targeted on high-value-added animal feed markets. The long research lead times, the high capital costs and still uncertain economic regime must deter any company considering SCP production. It remains to be seen whether Dansk Bioprotein will succeed in building its full-scale plant, never mind establishing Bioprotein as a traded product.

THE WORLD'S GREATEST PRODUCER OF SCP, THE FORMER SOVIET UNION

The Soviet Union directed an enormous increase in its SCP production capacity after the second world war and is now the world's largest producer of microbial protein. In 1985 there was over 1.1 million t/a capacity (Rimmington 1985) and the entire output is used as animal feed.

A researcher at Birmingham University, Anthony Rimmington, has made the Soviet SCP programme the subject of his PhD, and what follows draws on his work. According to Rimmington the origin of the state-directed expansion of SCP production was the coincidence of a shortage of livestock production with the problem of high paraffin wax oil in the Volga-Urals fields. The initial idea was to dewax the oil and produce much-needed animal feed in one process – the same conception that BP used at the start of its project.

The first plant of 1,500 t/a capacity was built in 1964 at Krasnodar. Rimmington (1985a) believes that the plant used a gas-oil substrate with separation of the yeast and feedstock after fermentation using solvents. Rimmington reports that the Soviets found the feed to be contaminated with polycyclic hydrocarbons, and that this persuaded them to build a second experimental plant at Ufa with a capacity of 12,000 t/a based on a purified n-alkane substrate. Toxicological testing apparently led to confidence in the safety of the n-alkane-based product and a series of giant plants were built in the succeeding years, all using purified paraffins as a substrate.

Ufa   100,000t/a, 1968
Gorki   200,000t/a, 1970
Kirishi 100,000t/a, 1972

Six other n-paraffin plants of similar capacity have probably been completed, while a further three with a combined capacity of 540,000 t/a were 'nearing completion' in 1985. These plants were built as part of the 1980 Soviet five-year plan, which had a target of 5 million tonnes capacity of SCP by 1985. This they failed to reach, possibly because of the belated attempt to install yeast dust scrubbing equipment in the exhaust gas flues of these plants. It is not clear why unambiguous announcements have not been made about when plants have opened and what percentage of capacity is being used.

The advent of *glasnost* allowed some of the problems with these plants to be discussed in the press and a key article in Pravda was summarised and reported in the *Guardian* by Martin Walker in 1988. This article reports the incredible furore over the Kirishi plant and the problems it is causing for the people in Kirishi town (population 60,000). According to Walker this article was 'the most appalling account of Soviet industrial pollution since the Chernobyl disaster' (Walker 1988). The article can be summarised as follows:

- The deaths of 12 children have been attributed to the Kirishi SCP plant emissions
- There are 100 permanent invalids in the town hospitals because of dust emissions
- The incidence of bronchial asthma is ten times higher than the Soviet average.
- About 4,000 people are suffering from 'degenerative' allergies linked to the plant.
- Some 12,000 people demonstrated against the plant on 1 June, 1987.
- A total of 300 defects in the plant were found by health and safety officials.

Pravda reported that over 2 tonnes of pollutant was settling on the town every day and that the scrubbers cleaning the factory exhaust system had been deliberately shut down while the plant continued to operate for a year and a half. Throughout this period demonstrations by the populace had continued with the effect that Kirishi was repeatedly visited by the Minister for Bioindustry and given promises that the pollution would stop. This ministry was originally given the task of planning SCP production and SCP remains the principle Soviet bioindustry.

A meeting of doctors from the surrounding region has condemned the plant while the authorities have responded in a confused manner. Those who have spoken against the plant have been repeatedly criticised and this article claimed that attempts had been made to silence protesters (Komsomolskaya Pravda 1988). The *Pravda* article also refers to 20-year-old toxicological tests on the plant's product, 'Paprine', which showed that second and third generations of animals fed large doses suffered from reduced fertility and disease immunity. Another report from a hospital authority described meat from animals fed on Paprine as undesirable for human beings, with stomach and bowel disorders common amongst people fed on such meat. *Pravda* also contains the cost of the attempts at lowering plant emissions – 500 million roubles, approximately £500 million, were spent in an effort to reduce emissions from this one plant by a factor of six. (Komsomolskaya Pravda 1988).

Pravda also reports a reluctance by farmers to use the protein concentrates from the plant, with the farming ministry of the Byelorussian republic banning their use as feedstock. Apparently thousands of animals have died from eating food yeasts from the Kirishi plant. A collective farm in the Leningrad area banned the use of 'BBK', another name for a form of the yeast feed, in spring 1987. The response of the plant authorities to this loss of consumption has been to direct their product to fur farms. Once again, there were animal deaths related to the yeast product and the Kirishi plant biochemists were sued, unsuccessfully, for £60,000 by the collective authorities of the fur farm.

These Soviet SCP ventures have frequently been criticised by Western companies like ICI who were involved in negotiations to license their SCP technology to the USSR. Rimmington refers to an 'informed source' in ICI that described meat that had come from animals fed on Soviet SCP as having 'an unpleasant taste and smell' (Rimmington 1985: 37). Rimmington has also found references to Soviet scientists worried about the possibility of large-scale epidemics triggered by the venting of yeast-laden dust from these plants. He reports that dust-catching devices with capacities of 1,150,000 cubic metres air/hour were being installed in 1985.

There appears to have been a debate over whether the levels of carcinogen in the n-alkane SCP product should preclude its use as a human food supplement. Rimmington (1985a) reports that the Soviet Institute of Nutrition has so far refused to allow its use in human food, although the producers of SCP have tried to introduce it.

The reaction of a microbiologist in one of the Western developers of SCP plant reacted strongly to the Soviet situation: 'My God! They must be doing something bloody awful – totally incompetent'. But both BP and ICI spent time negotiating with the Russians over licensing their technology and formed a negative impression of Soviet technological capabilities:

> They didn't seem to be able to do simple things like pour out large slabs of concrete . . . they seemed to be needing inputs in some astonishingly simple things. Like making decent stainless steel . . .

> apparently they couldn't do it . . . one has doubts about their ability to operate.

There is other evidence that the Soviets have had trouble with the operation of their production technology. Rimmington (1984) collects various reports of Soviet approaches to Western equipment manufacturers where the nature of the purchase implies they are not satisfied with indigenous equipment in their SCP plants, in particular centrifuging and separation equipment.

One manager suggested of the BP negotiations that the Soviets exploited BP for technical information they should have been buying.

> All the Russians kept doing was saying, we're very interested in your SCP technology, it will require a bid and a process design from you . . . so we designed a number of process plants for the Russians. And every time it wasn't quite what they wanted, so there was a redesign. At the end of the day the Russians had got chapter and verse on the SCP plant, supplied by BP. Three years later they announce to the world that they are fermenting paraffins at a million tonnes a year by a process that is remarkably similar to the BP process.

But this manager was as contemptuous as others of the Russians ability to operate such plant:

> It's not BP technology . . . it's BP style of engineering and fermentation. It's typical of massive Russian industry, it's no different, just as inefficient as the rest of their industry . . . whether they'll ever publish efficiency data is another thing . . . if they only got one gram of yeast out of it they would say they had a process for making yeast.

In the early 1980s the Soviet planners were concerned that they should develop SCP on alternative substrates, as the supplies of n-paraffin were thought to be limited. There was a concentration of attention on methanol and natural gas, with the first experimental plant using natural gas completed in 1983. Rimmington refers to *Pravda* articles talking of nearly 1980 targets of building many plants of 'unprecedented' capacity, of 350,000 t/a to 500,000 t/a, based on the new substrates.

It was this interest in developing natural gas as a feedstock that led to the approach to ICI to negotiate over the transfer of Pruteen production technology. Kostandov, the then deputy prime minister, visited the ICI plant at Billingham and Harvey-Jones had talks with

Tikhonov, the prime minister, in Moscow, but no deals were made. One of the ICI plant construction managers thought that this was because the Russians had developed their own technology for fermenting natural gas to protein, similar to the Shell process, and he too had nothing favourable to say about Soviet engineering abilities:

> Now, I don't want to be anywhere near their plant when they try to start it up. You can go on and on about Russia and the problems they have . . . because of inefficiencies of one sort or another.

Other ICI managers criticised the 'primitive technology' the Soviets were using and one commented that they were . . . economically grotesque, very low methane efficiency, they want cheap capital costs, not the most modern technology'. The Soviets have reconstructed one of the Ufa plants to operate on ethanol at 100,000 t/a SCP capacity. Other important SCP processes are SCP from hydrolysed wood chips (500,000 t/a) and SCP from sulphite waste liquor (140,000 t/a) (Rimmington 1984). The Russians' experimental substrates have included rice husks, corn cobs, cotton hulls and bagasse.

Rimmington has suggested in all his articles that the Soviets were about to launch an expansive programme of plant building, possibly based on methanol or natural gas. For example,

> A massive increase in SCP production can now be expected over the next ten years. N-paraffin plants currently being constructed will come on stream, new factories will be built to utilise natural gas as substrate and it is likely that methanol-utilising plants will also be purchased from ICI in the near future. Undoubtedly, if these developments take place the Soviets' claim to be world leaders in industrial biotechnology would be strengthened. And perhaps more importantly, Soviet purchases of grain from north America might be curtailed.
> (Rimmington 1985b: 37)

From the perspective of 1992 this comment certainly reflects the overoptimism of the Soviet central planners.

The economics of n-alkane SCP production appear to be as marginal in the Soviet Union as they have been elsewhere, a typical comment from a Western manager being, 'for some socialist countries economics doesn't mean anything'. The reason for continuing construction of such plants must be a political one intended to reduce dependence on Western grain supplies for animal feed and to avoid tackling the problem of inefficiencies in distributing Russian-grown grain. However, Rimmington also shows by reference to the Soviets own figures for animal feed consumption requirements that at 1.5 million t/a, SCP would be only a small contribution towards the Soviet needs to feed more than 250 million animals. There are also figures which suggest that only 85 per cent of the raw protein requirement of Soviet animals is being supplied, and that this is a situation which has persisted for over 15 years (Rimmington 1984: 8). As one manager commented,

> There is a terrible lack of protein for animal feed in the Soviet Union. It's still the case, they use cotton feed meal of bad quality and they have always used much high protein feed of poor quality.

It is no wonder that the planners wanted to increase SCP output dramatically and the Soviets were interested in enormous contracts for plant construction Liquichimica were asked to put in a bid for ten 100,000 t/a plants.

The Soviet Union has shown that SCP plant can release air pollutants that damage health and that the product can become toxic if production and distribution is not handled well. The inability of the Kirishi population to close the plant when it continued production in order to meet planned targets, irrespective of the air pollution it was causing, is a feature of the Soviet political system – in the West negative public opinion led to the political sabotaging of the Japanese and Italian SCP plants before production ever started, and in the face of evidence that the plants would not pollute and that the products were safe.

SCP IN JAPAN

Once BP announced that it was researching the growth of yeast on hydrocarbons Japanese companies were quick to follow similar research avenues. At one time there were more than 12 companies researching and developing SCP processes, but as the pioneers ran into problems most of these quietly dropped SCP. The debacle in Japan over the alleged toxicity of SCP contributed to the problems of BP and Liquichimica in Italy by arousing suspicions over the toxicity of the product. But the attitude towards multinational companies in Italy was close to that in Japan at this time – the public already did not trust either government or companies, so when these tried to show that SCP was a safe product they were assumed to be lying, in the interests of profits, by a significant proportion of the media and public.

The pioneers were Dainippon Ink and Chemicals and Kanegafuchi

Chemical Corporation. As in Italy, public reaction against SCP reached a pitch when companies began building full-scale plant. After 12 years of research Dainippon began laying the foundations for their 60,000 t/a plant, which was to be completed in 1974 and had an option to increase output to 120,000 t/a at a later date (European Chemical News 1973). Kanegafuchi had similar plans and had even tied up an agreement with the Japan Association of Agricultural Cooperatives to part finance their plant and to buy the bulk of its production (Chemical and Engineering News 1973).

According to the contemporary Chemical and Engineering news reports, the official company reasons for the suspension of their plans was 'vocal opposition from consumer groups' (Chemical and Engineering News 1973: 9). The same article reports the companies' private comments that they were given 'administrative guidance' to drop the protein projects by the Japanese Ministry of Health and Welfare, which in Japan amounted to an order to do so.

In December 1972 the Health Ministry approved the results of toxicological tests performed by the two companies on samples of their protein products and cleared n-alkane derived proteins as non-toxic when used as animal feed. With this approval and the beginning of plant construction the campaign against the SCP plant really began.

The consumer groups who campaigned against the plants called themselves the 'Tokyo Liason Council' and the 'Liason Council to Ban Petroleum Protein'. They accused the Ministry of Health of accepting data submitted by the companies rather than commissioning its own studies. This was despite much of the toxicological research that the companies commissioned having been carried out by Japanese university medical departments, rather than directly by the companies. Clearly these groups did not respect the supposed independence of academic research if financed by companies with a profit motive. The other charge was that low concentrations of toxic material, originally present in the hydrocarbon substrates, might accumulate in the tissues of animals fed on these yeasts, to a level where they would pose a significant threat to human health (the same argument was used in Italy against BP and Liquichimica).

The response of the Health Ministry was to duck the issue by announcing that since 'petro-proteins' were an animal feed and not a human food, they were after all outside its jurisdiction. Shortly after this announcement in February 1973 the Minister of Health directed his staff to prepare a new set of regulations on the safety of petroleum-derived protein products; these regulations on the testing procedures required for SCP were designed to take years to complete, and so immediately after this decision came the advice to the companies to suspend their construction plans.[1]

Dainippon and Kanegafuchi both claimed to be continuing product development and to have complete faith in the safety of SCP at this time. Dainippon announced that it would continue to promote licensing arrangements overseas and to research alternative feedstocks and uses, such as methanol-based SCP and the use of SCP as a petfood, but continuing pressure on the government by public protest groups resulted in the Ministry for International Trade and Industry banning any new licensing agreements in April 1973.

Japanese government safety tests on SCP continued through the 1970s, with one programme involving the study of more than 5,600 hens fed on SCP over five generations. These testing programmes resulted in at least two major Japanese government clearances of SCP as a safe product in the 1980s – far too late to be of any use to the Japanese producer companies, or BP and ICI.

There were other products that were withdrawn in the early 1970s after similar forms of public organisation and pressure on the authorities over safety status (Arima 1979). What had happened was that public distrust of government and industry built up after a series of appalling industrial poisonings and pollution incidents, throughout the late 1960s and early 1970s. Probably the most important was the Minimata disaster. This mercury poisoning disaster involved illegal emissions of mercury into a bay and the wholesale poisoning of people in local fishing villages. The company responsible refused to admit the poisoning was occurring, then refused to admit its own responsibility. The government was seen to be tardy in prosecuting, lenient when it did so, and the company remained reluctant to pay compensation. The public learnt not to believe official company statements or rely on the government to safeguard the public interest. SCP was almost certainly a victim of this poor relationship between industry and public. The good that has come out of these experiences is that companies are aware that they can all be lumped together and 'tarred with the same brush' by one company's bad practices, and that active maintenance of image with the public is important – it can affect profits. Ironically, although politics eclipsed the Japanese forays into SCP, this may have been to their advantage since Dainippon would have found itself in the same position as BP in 1974 – a large-scale plant that had no chance of repaying the capital sunk in its construction. Box A.1 provides a summary of SCP developments in Japan.

Box A.1 Summary of SCP development events in Japan

1962–3	Basic research begun by Kanegafuchi Chemical Industry Co., Dainippon Ink and Chemicals Inc. and Kyowa Hakko Kogyo Co.
1968	Ministry of Science and Technology submits a report to the government advising of the importance of SCPs.
1967–8	Industrial production research begun by Kanegafuchi and Dainippon.
1968–9	Commercialisation research and feeding, safety trials begun by Kanegafuchi and Dainippon.
Oct. 1968	Kyowa Hakko Kogyo licenses BP technology with intention of building an SCP plant.
June 1969	Kanegafuchi signs feed supply contract with National Federation of Agricultural Cooperatives.
Aug. 1969	National papers report nature of health ministry's safety concerns, i.e. the possible presence of mycotoxins and carcinogenic alicyclic hydrocarbons in the three Japanese SCP products.
Oct. 1969	Special Committee on Petro-protein established by health ministry to investigate safety of SCP, toxicological tests begin.
1969–71	Ministry of Agriculture sponsors nutrition trials for SCP as a fish feed, eventually announcing that SCP is a safe and nutritious food for fish.
Dec. 1972	Ministry of health announces that Kanegafuchi and Dainippon products are safe if used for animal feed.
Dec. 72	Protest movements begin to organise.
Jan. 1972	Opposition parties raise health concerns over SCP in Japanese Diet.
Feb. 1973	'Voluntary' abandonment of SCP commercialisation by Kanegafuchi, and later by Dainippon on grounds that investment was too risky given furore over safety. Both companies continued to insist products were safe. Campaign continues against SCP.
Apr. 1973	Export of SCP technology prohibited by MITI, excluding established agreements, i.e. Kanegafuchi has licensed its technology to Liquichimica and Dainippon has aided the construction of a plant in Romania.

Source: Arima 1979.

SCP FROM WASTES

The only surviving SCP production in the West today is based on the fermentation of the high carbohydrate wastes which are a by-product of wood pulp, dairy and confectionery manufacture. Like sewage waste, these wastes can be dumped in the environment where uncontrolled microbial digestion will break them down into carbon dioxide and water. As in the case of sewage wastes, most countries regulate such dumping, although not because of the risk of disease, but because of the biological oxygen demand (BOD) of the process of microbial digestion. Microbial digestion removes the oxygen dissolved in river water (where the wastes are usually dumped) and, if there is a sufficient concentration of waste, can consume oxygen faster than oxygen dissolves into the water from the air. If this happens, animal life dies and anaerobic microbial action begins, which produces toxic compounds such as hydrogen sulphide that poison the water further. Water companies, responsible for maintaining the health of rivers, will generally charge companies by the BOD value of their wastes. So these wastes have a 'negative' economic value to the companies, which is why SCP production can be of interest to companies. Controlled fermentation not only reduces the BOD value and so dumping charges for liquid wastes, but the yeast biomass that is produced has a positive economic value as an animal feed.

Wood pulp waste fermentation

Wood is chemically composed of a complex carbohydrate called lignocellulose, which in one review of the potential to grow SCP on wastes has been called 'the world's largest reserve supply of renewable carbohydrate' (Steinkraus 1980: 137). Lignocellulose is basically wood, and the by-products of pulp processing for paper production offer the largest potential of any waste carbohydrate material for SCP production. When paper pulp is made by dissolving wood chips in sulphuric acid, only the cellulose carbohydrate is used, the lignin composing the other half of total wood carbohydrate ends up in the form of lignin sulphate dissolved in water, and sugars. This by-product, or waste, is called 'sulphite liquor' and contains 28 per cent sugars by weight. Since they are a by-product of the continuous process of making pulp, lignocellulosic wastes have the advantage that they are consistent in quality and are produced at a constant rate, making a consistent quality SCP product possible. In Scandinavia it is the major tonnage carbohydrate waste and it was here that the most important SCP process to be based on lignocellulosic wastes was developed, the Pekilo process.

The Pekilo process was developed by the Finnish pulp industry's collectively owned R & D institute in Helsinki. Of the 26 companies that belonged to the institute, six that owned sulphite mills decided to support research into the Pekilo process in 1963–4, with development beginning in 1968. The head of the institute thought the incentive to use the process was greatest in the 1960s when Finnish environmental legislation was tightening rapidly and companies did not know how stringent regulations would be in just a few years time. The trend for greater legislative severity and the trend for world protein prices to rise made the case for the research and development of the Pekilo process, but there was also some idealism involved at the Pulp and Paper Institute:

> We thought, why should wood be converted only to toilet paper, what about food for the starving? And at that time, there was lots of idealism. The disaster in Peru, 1972, the anchovy disaster – protein prices were sky high. It boosted activities here. United Paper Mills [later merged with the company Metsa Selva] built the first plant on the basis of these high protein prices.

Of the two Pekilo plants that were installed, only the one owned by the Metsa Selva company is still in operation at Jamsankoski, Finland, where it produces 8,000 t/a yeast fodder in a two-stage fermentation process. The first stage produced ethanol by anaerobic fermentation of the hexose sugars in the liquor and the second stage consumed the pentose sugars and produced SCP. According to a plant manager, the only reason for the second, SCP stage was a need to lower the BOD of the effluent which the plant discharged.

> Since the 1970s we have become highly BOD conscious, legislation is passed daily here . . . there is no salmon in our lakes any more, and acetic acid is one component of this problem . . . The water authorities are breathing down your neck all the time.

Table A2 Economics of Pekilo protein production

Item	Operating cost ($/tonne)
Value of Pekilo protein	110
Cost for loss of heat value of liquor to pulp mill plus power and steam consumed in process	50
Nutrients	18
Utilities	11
Labour	11
Maintenance and packaging	20
Total costs	110

Source: Romantschuk 1975: 348
Note: Power consumption 1,250 kWhr/tonne

The yeast fermentation had the advantage that it consumed the acetic acid in the effluent which made a disproportionate contribution to the BOD value.

The value of the yeast protein that the plant produces is set by the prevailing soya price (which represents the price of bulk protein) and a simple economic calculation shows that it just covers operating costs (Romantschuk 1975), but there is no hope of repaying development costs or capital expenditure on the plant. Unfortunately, protein prices had fallen back again by 1975, ruining the economic forecasts which had justified the building of this plant, while a rival pulp-producing technology emerged as more economic than the sulphite process, called the Kraft process. In the rival Kraft process the sulphite liquor is evaporated to a dry waste and then burnt to supply more than enough energy to run the plant, whereas the Pekilo process is a net consumer of energy – the upshot is that in a price regime of high energy costs and cheap protein prices it is more economic to burn the sulphite liquor and avoid the capital expenditure and operating costs of a Pekilo plant (Table A.2).

Yeast fermentation can be run aerobically, which maximises the production of yeast cell mass, or anaerobically, when alcohol is the major product. The manager of a Norwegian sulphite plant owned by the company Boregaard, which had installed an alcohol fermentation plant before the Second World War, commented on the shifting economic climate that might have made SCP production profitable: 'Every time the ethanol price tumbles we have considered using some kind of SCP process'. Boregaard produce 16,000 t/a ethanol using a Saccharomyces yeast and about 1,000 t/a of fodder yeast. The capital cost of adding an SCP plant for this company would have been relatively cheap, but even when the price of ethanol was especially low, feasibility studies in 1950, 1973 and 1979 showed that it was not economic to manufacture SCP. As with the Pekilo process, the manager of this plant did not believe the alcohol fermentation plant would be economic to install now, given the capital costs, and Boregaard's experience of the economics is probably representative of the majority of carbohydrate waste producers in Europe, where fermentation was used simply to produce bulk protein for animal feeds.

However, yeast fermentation of sulphite liquor selectively removed the sugars from the waste, leaving lignin sulphate. Boregaard used their fermentation to 'refine' the liquor in this way and to sell the resulting lignin sulphate syrup as a specialist lubricant for oil-well drilling bits. This specialist use made it worthwhile for Boregaard to continue operating their fermentation plant.

> We are a combination of a pulp and chemicals firm and this is a
> fairly unusual combination. I think with Metsa Selva in Tampella,
> Finland, we are the only two companies in Europe with such a
> combination of interests . . . It's true that it's because of our
> chemical operations that we are interested in removing sugars
> from lignose sulphates.

So the unusual structure of the company meant that an additional
use for the ethanol fermentation plant could be found that ensured
its survival. The company which operates the Pekilo plant, Metsa
Selva, is also part of a chemical combine and managers also gave the
availability of chemical expertise and chemical markets as a reason
for their venture into SCP. But when asked why Metsa Selva built a
SCP plant when Boregaard were deciding not to, a plant manager
replied that

> I'm not sure how much I can disclose, but the protein market was
> somewhat protected in Finland, in that there were some limits on
> the consumption of imported soya feeds. The price was higher
> than in the EEC, and this made a significant difference to the
> economics. Also the Norwegians [Boregaard was a Norwegian
> company] had their fishing industry and cheap fish-meal [which
> would depress protein prices].

So economic conditions differed sufficiently that the two companies
took different decisions on the viability of SCP. As the Kraft process
continues to replace sulphite mills in Finland, even the operation of
the remaining Pekilo plant must be in doubt.

The Kraft process neatly avoids the problem of water pollution by
burning the evaporated sulphite liquor – but incineration produces
the acid air pollutant, sulphur dioxide. In sparsely populated Finland
this may not matter, but in central Western Europe there exists one
of the toughest air pollution regulatory regimes in the world, and
such emissions are not possible. The result is that there are no Kraft
mills in either Germany or Switzerland. Whereas Finland has a
problem with water pollution, Switzerland and Germany also have a
problem with sulphur dioxide pollution and acid rain. The trend to
tougher regulatory control has created a selection environment in
which existing sulphite mills must continually invest in pollution
control equipment just to stay in business. Most mills have been
closed down completely, but in Switzerland there is one remaining
mill, run by the Cellulose Attisholz company in the Canton of Berne,
which has survived by cleverly anticipating tightening regulatory

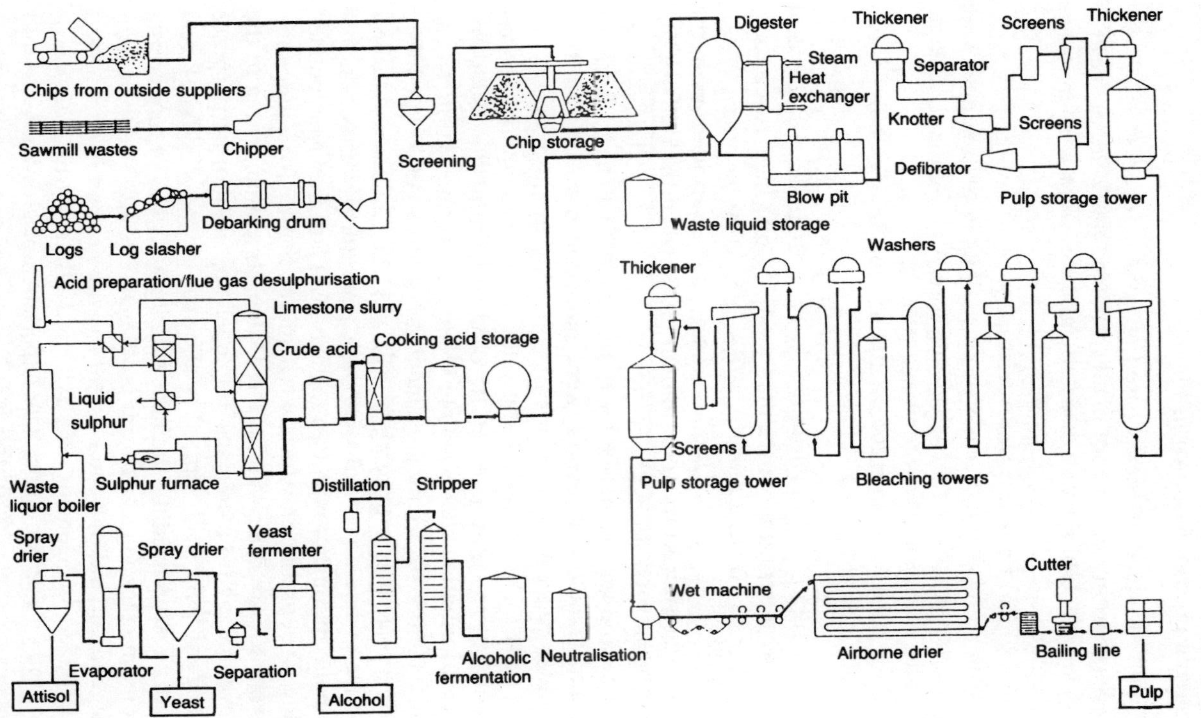

Figure A.1 Flowsheet of the Cellulose Attisholz pulp-processing plant
Source: Cellulose Attisholz 1977

control. The following represents a list of this plant's major BOD reduction systems since it was built in 1881 (a diagram of the pulp-making process is shown in Figure A.1):

1881   Pulp production begins based on sulphite process.
1914   Saccharomyces ethanol plant installed, waste liquor burnt.
1947   Candida Utilis strain used to produce yeast fodder, waste liquor burnt.
1970s A 220,000 t/a waste-water treatment plant installed.
1976   Human food grade yeasts produced on alternating basis with fodder yeast, as part of a long-term development of high-value-added markets for yeast protein.
1986   Pilot plant built to recover sulphur dioxide from burnt waste emissions.

Until the ethanol plant was installed, in 1914, there was 100 per cent discharge, with pollution worse than today despite increases in pulp production from 30,000 t/a in the 1920s to 125,000 t/a today.

The installation of the fodder yeast plant coincided with the immediate post-war shortage of animal feeds, when the real price of feed made such an installation economic. As with the Pekilo process, two fermentations were necessary to remove all the sugars from the liquor, and Cellulose Attisholz were pioneers in developing their yeast fodder process as a means of removing pentose sugars and lowering the BOD of effluent. The yeast fodder and ethanol process were sufficient to control the BOD of effluent until the 1970s and company information sheets claimed that yeast fermentation was an economic means of treating effluent as late as 1977 (Cellulose Attisholz 1977). It is intriguing to ask why the Cellulose Attisholz mill survived while dozens of rival mills went out of business rather than respond to tougher pollution regulations. Plant managers pointed to the R & D department, which was a unique feature for a pulp mill. There were 40 researchers in the R & D department compared to the usual seven or eight technicians employed for monitoring purposes in a standard pulp mill, and this allowed the possibility of real exploratory research rather than the simpler activities of monitoring operations and tackling short-term problems. The R & D department had designed and built all the effluent treatment plant itself and had even succeeded in licensing the SCP and effluent treatment technology to other sulphite mills. The experience of growing yeast and the engineering expertise of the R & D department meant the plant management could be critical of the design solutions on offer from engineering companies for the waste water treatment plant.

. . . the engineering companies were marketing poor plant. They were not designed to cope with the requirements of micro-organisms . . . they were engineering companies with a poor appreciation of the chemical and biological processes their plant was supposed to be catering for. So, using our knowledge of the characteristics of yeast grown on waste, we designed the plant ourselves . . . it was twice as efficient [at lowering the BOD] and half the price of our competitors[2]

This treatment plant supplies 25 per cent of the plant's electricity and 60 per cent of the energy requirements of an adjacent waste incinerator plant built to take the excess steam. The company patented the design and sold between 100 and 200 similar units for sewage treatment.

Box A.2 Principal stages of the Cellulose Attisholz yeast production process

1 Thermal pretreatment of spent sulphite liquor in degasifying column.
2 Neutralisation and addition of yeast nutrients.
3 Air lift fermentation with central vat surrounded by circulatory tubes for air injection, (homogenises mixture, cools, oxygenates and allows carbon dioxide degassing).
4 Continuous overflow of yeast from fermenter.
5 Separation.
6 Multi-stage washing.
7 2 stage evaporation plant, thermolysis and pasteurisation.
8 Spray drying of yeast cream (22 per cent solids) to marketable powder, 7,000 t/a.

Source: Cellulose Attisholz 1977.

Although the plant management initiated research on the waste-water treatment plant because they anticipated tightening of pollution regulations, it was their belief that by selling versions of their design to other sulphite mills they had threatened the established business of the engineering companies they had found so unsatisfactory. These companies had then complained about the emissions of the Cellulose Attisholz plant to the Cantonal authorities, who then gave the plant four years to decrease their effluent BOD, or the plant would be closed. The only reason they had been able to comply was because they were at an advanced stage with their own research, and this lesson had had the effect of supporting their attitude to regulation:

> The trend is to become stricter in time . . . [so] if you meet
> one set [of regulations] exactly, you are soon asked to improve
> on them, [although] Switzerland has the strictest regulatory
> environment in Europe.

The company aim to do better than the current rules demand,[3] and
to this end they have begun work on the development of a non-
sulphur, non-chlorine pulping process, which they consider the only
option for new pulp plant in Switzerland. The entire strategy of the
R & D department is dictated by intelligent anticipation of legislation
and it has now decided that this requires the abandonment of the
core processing technology. The plant management clearly feel
vulnerable – before the 20 million Swiss francs desulphurisation plant
was built, Cellulose Attisholz produced 10 per cent of total Swiss
sulphur dioxide emissions.

The company was lucky to have as founder a chemistry PhD, Dr
Sieber, who had established the tradition of an active and pioneering
R & D department. Although the company was no longer family
controlled, the R & D department had earned high status as the
saviour of the plant and so management had never considered the
short-term measure of cutting costs by closing R & D. Instead they
were relying on the department for their future. As in Scandinavia,
the Swiss managers thought that, despite being economic at the time
they were installed, neither the yeast nor the ethanol plants would
be viable investments under existing economic and technical conditions. But since the capital investment had been written off in
accounting terms and the plant was operationally profitable, it
remained in use, a physical reminder of past environmental conditions.

Protein from dairy wastes

The move to replace the core pulp-processing technology is a radical
step, but so is the effort to build a market for the yeast protein as
human food, especially since food marketing is outside the experience of a wood pulp manufacturer. This step was much easier to take
for the French dairy products company Bel Industrie, which produce
18,000 t/a of 'lactoserum', the waste product which remains at the
end of butter and cheese manufacture. Once again this volume of
waste presents a disposal problem that yeast fermentation can solve,
but as one of the largest-scale dairy producers in the world Bel
Industrie have a unique problem – they believe they have not
succeeded in marketing their fermentation process to other dairies,

because they are too small to make the capital investment worthwhile. Bel worked with the Association des Souches, the French
Yeast Association, to develop a continuous fermentation process that
used a mixed yeast culture (Box A.3). Three yeast strains worked
together to consume the separate components of lactoserum and the
alcohols the yeasts themselves produced. Over the years a stream of
minor process innovations and adjustments raised the yield from
42 per cent to 53 per cent, (yield is the ratio of the weight of yeast
produced per unit weight of solid substrate). Three yeast strains
worked together – lactoserum (lactose and lactic acid) and the
alcohols. Yeast production requires enormous amounts of energy for
the aeration, drying and separation processes, so the 1973 oil price
rise led to a tripling in the cost of producing the yeast, leaving Bel
unable to compete in their animal feed market. This prompted Bel
to begin a search for higher-value-added markets, and in 1978–9 they
developed a product called 'Sassiyen', which was a calf milk replacer
rather than a crude yeast fodder. They also began to sell yeast into
the dairy-free food market, where it could be substituted for milk
products in human food.

Bel worked with Montpelier University to show that when the
culture was contaminated with a foreign yeast strain, then fermented
continuously, the same stable mixed culture was always the result; in
other words, it was resistant to toxic infection and the quality was
reliable. They sought this independent seal of approval to help them

Box A.3 Development of Bel Industrie's food yeast production process

1955	Research into strains begun in collaboration with Professor Keilling of the Laboratoire National Agronomique de Paris, (LNAP).
1956	Pilot plant built at Vendome and already producing 300 t/a of yeast by continuous fermentation.
1966	Main plant built at Vendome of 1,500 t/a capacity; it consumes the entire output of serum from Vendome dairy and produces cattle fodder and protein concentrates. Yeast is flocculated and filtered in harvesting process.
1964–74	Research shows yeast can be used as protein and vitamin supplement in human food, although 90 per cent of output is going to the animal feed market.
1974	Plant built at Sable sur Sarthe. Ultra-filtration process incorporated to replace flocculating process. Protein is now soluble and commands a higher price and there are more human food uses.

actively market their yeast product as a human food, which would add even greater value to the yeast than when sold into the calf milk replacer market. They now have two factories, one producing Sassiyen and one producing human food, and both the animal and human food markets became increasingly profitable in the late 1980s.

After the shock loss of the animal feed market in 1973, Bel have actively developed a series of small markets of higher value which they believe will make them safe from future sudden shifts in oil price. Bel are prepared to leave some of their yeast production capacity of between 10,000 and 12,000 t/a unused while they take the time to develop these higher-value markets. In all cases they sell their yeast to other companies for further processing, and they have kept individual market size down to around 1,000 t/a, despite offers to buy supplies of yeast at 3,000 t/a.

Bel examined the domestic animal food market and found that there was a small, profitable niche for yeast as a vitamin supplement for improving dog coats. They also have a market for fish feed, which is important in Germany. The technical director thought that both these markets had high potential value in the future, but that development partly depended on Bel demonstrating the safety of their product. Other openings that Bel are actively considering are aquaculture of fish and prawns in Japan.

Bel judged the human food market to be largest in the US, then in the UK. However, the UK was the most difficult market because of the vast quantities of surplus brewery yeast that was produced and here Bel were not well known. The US market was worth 33,000 t/a, but the problem with all the human food uses was that there were already competitor products in place. Bel also had the problem that lactic yeast was relatively unknown and had to win acceptance. The major problem was to find markets where its different taste would be a positive advantage – the difference between lactic and brewer's yeast are small but important. The organisms were the same, but the production process changed the quality of the yeast proteins and lactic yeast was the only yeast product to contain vitamin C. On the other hand, brewer's yeast contained slightly more vitamins B1 and B2.

Their most profitable market is in producing dairy-free products for human consumption and the most developed market is that for dairy-free chocolate, sold in Germany, France and the US. The high-quality powdered yeast is used as a milk replacement. Another human food market is as a vitamin supplement and Bel sell their yeast to be turned into tablets for child diet supplements, although here there is much competition from breweries and bakeries selling their waste yeast mass. However, lactic yeast has the advantage compared to brewery yeast that it has a sweet rather than a bitter taste, because of the lactose substrate on which it grows. Another human food application is in charcuterie, where the water-absorbent qualities of the yeast can be used to bind proteins in sausages; again lactic yeast has a competitive advantage because of the presence of water-absorbing galactose in the yeast cell walls, which brewers yeast lacks.

For all these yeast processes there has been an unsteady retreat from profitability, and in the companies described here management efforts to change the process and market were stimulated by regulatory or economic shocks and insecurity. The technology appears as a rather vulnerable creature which is being evolved by its management in an attempt to keep it alive.

SCP FOR THE THIRD WORLD

Most SCP projects by number were designed to produce protein in the third world and were commonly known as 'village technology'. These projects aimed to ferment some common local source of carbohydrate and so improve its nutritional value, usually for the local animals.

SugarCo never believed in the idea of a protein gap during the 1960s, but believed instead that, while the First World would remain in protein surplus, the problem was increasing the amount of protein used in the poorest tropical countries of the Third World. The argument was that the diet of the people who live in the 15 degrees band either side of the equator is protein-short, because the plants they eat do not produce either enough protein or the right sort, while their food contains too much carbohydrate or too much fat. The attraction of village technology was that it was an alternative to the use of fertilisers and the intensification of conventional agriculture which required First World technology in the form of imports. By feeding animals on SCP made from a local fermentation, more protein would be made available for people. The projects were always technologically simple: 'we thought it was necessary not to have anything too sophisticated because in developing countries you can't have PhDs and white coats running around setting these things up'. SugarCo always believed that selling their technology would depend on political support and that another company would be unlikely to buy it without such support. Unaided private capital was never really interested in these projects, because the returns on their use were in terms of improving the diet of people on the margins of money economies and at best they reduced pollution and could save foreign currency by substituting for soya-meal imports. Only a government might find these effects worthwhile and SugarCo developed their SCP technology with the hope that they could convince a government to buy it. The market for these technologies would have to be political and it was the retrospective judgement of some of the SugarCo managers that their research effort lacked an adequate marketing input.

The carob project

A candidate country for this type of technology was Cyprus – a country with a protein problem, because very little grass can be grown and animals have to be fed on imported feeds. The carob tree is native to the Mediterranean area, and although carob beans were collected, the husks were simply thrown away. One of the SugarCo R & D staff found that these husks were 50 per cent sucrose and the company devised a plant which would use Aspergillus Niger to ferment the sucrose to provide a high-protein animal feed.

In the early 1970s the company began negotiating to sell the SCP process to the Cypriot government headed by Prime Minister Makarios. SugarCo managers felt that the government were very keen, but despite the hopeful beginning no progress was ever made. One of the SugarCo consultants at the time explained this as a result of the first of the little wars that ended with the Turkish invasion, where the urgency of this situation meant that secondary issues like SCP were abandoned. On the other hand, the war ended in stalemate more than ten years ago and if the carob project had really been economic it could have been introduced since. One reason the company gave was that as soon as SugarCo ordered husks the price rose sharply as producers realised that what had been thought of as waste, actually had an economic value, which upset SugarCo's economic evaluations. However, although the husks were waste for the bean harvesters, pigs would feed on them quite happily. So the carob fermentation was upgrading husks which were already being used as animal feed, and although the pigs needed some protein supplement to a diet of husks, there were alternative ways of supplying this, for example, by simply adding urea to their diet. There was no need for the intervening and labour-intensive step of collecting them, inoculating and fermenting the husks before feeding them to the animals.

Citrus waste in Belize

SugarCo devised another SCP process to upgrade citrus waste into animal feed, again with Aspergillus Niger. They even proceeded to build a pilot plant next to a citrus canning factory in Belize, although the reason for siting the plant in Belize was political. The Belize project was a version of a process developed for the fermentation of molasses. An econometric analysis of the original process showed that the process was only economic when there was a true waste, one which you would be paid to take away, which was the case with citrus waste. However, citrus waste was difficult to ferment, while the Belize plant was isolated and unable to distribute its product effectively. Once the pilot plant was built, the canning factory lowered the sum they paid SugarCo for treating their waste and the economics worsened as a result. One of the managers said that even without this, it had not been viable, and indeed, after three or four years SugarCo closed the plant.

The only market idea for the plant came after the decision had been taken to build it. SugarCo found communities of Mnemonite farmers in Belize and these farmers traditionally raised chickens and would probably still wish to do so. Since there was no chicken feed in tropical Belize because grain would not grow and imports were too expensive, it was thought that these farmers could have provided a market for the upgraded citrus waste – in the event it was not a market that supported the operation of the plant. A current manager commented on the market for these projects:

> It was seen but not enough attention was paid to it. It was assumed that we could encourage people to take up more intensive feeding of their animals. And that is quite a difficult thing to do, to introduce a new product and develop a new market for it at the same time. Very difficult.

On both technical and market grounds, Belize was a poor choice.

The failure of village technology to find markets

Managers in the companies that pioneered large-scale SCP plant were scathing on the subject of village technology. Referring to the 'daftness' of village technology, one microbiologist argued that this kind of fermentation could only be done by really high technology, otherwise the protein was not safe for the animals to eat. It was too easy to contaminate the protein with foreign micro-organisms during a low-technology fermentation. Another commented that

> There have been certainly hundreds of projects of this kind . . .
> hardly any of them going now, hardly any got beyond the pilot
> stages. Two basic reasons: one, they didn't work because you
> couldn't get a safe and consistent product. Two, the markets were
> not there. Western scientists assumed humans would eat it or
> animals would be around to feed it to. Both assumptions were
> wrong.

The complex feed-compounding industry in Europe had no equivalent in the poor countries for which SugarCo had developed its SCP projects. In these countries the animals feed themselves by scavenging and are not sufficiently concentrated to justify delivery of animal feed from a capital-intensive plant operating with scale economies, nor is there the road infrastructure to make those deliveries. The lack of economic specialisation and then concentration of production make waste-based protein projects uneconomic. There was also the issue that there were better ways for poor countries to spend their limited resources than on soya-meal import substitution.

Where SCP was produced from wastes in the industrialised countries it was because legislation forced companies not to pollute the environment. But as one manager commented, in the Third World

> no one gives a damn about pollution, the problem is eating
> tomorrow . . . so if you happen to be sticking something in the
> water that kills all the fish, they really don't care. Third World
> governments only have time to pay lip-service to that.

Then there were the problems of having Third World populations accept a new source of food, even if it could be produced economically. Cowen illustrated the point with the story of how some years ago millions of tonnes of wheat were sent to India to alleviate a famine, but were left to rot in warehouses because, despite starving, the people would not eat a non-familiar food.

It is ironic that all of these projects depended on the idea of an insufficiency of protein, based on the then current understanding of nutrition science. The idea that people could become protein deficient rather than simply starve for lack of food has since been dropped by nutritionists, but as an element of market thinking it helped to drive the development of these hundreds of village technology projects.

Glossary

Alkane	Any saturated aliphatic hydrocarbon
Cause map	Weick's (1979) term for the retained, linked cognitions which form an individual's set of casual relationships
Cognition	Knowledge that is comprised of elements of perception
Cognitive mapping	A process of constructing a mental map of one's environment
Comparative analysis	The examination of similarities and differences based on the categorisation of the text by the researcher
Empirical	Relating to experiment and observation rather than to theory, based on practical experience
Enactment	The process by which certain cognitions are selected and then retained into the causal map, i.e. they become part of an individual's experience
Fermentation	The growth of micro-organisms on an organic substrate which they metabolise
Grounded theory	The process of the discovery of theory from data using a method of comparative analysis
Hydrocarbon	Any chemical consisting of hydrogen and carbon alone, e.g. alkanes
Innovation	A novel product or device or idea, not necessarily commercially successful
Methane	The simplest alkane and the main constituent of natural gas
Methanol	The simplest alcohol, the first stable oxidation state of methane, highly soluble in water and toxic

Network	The set of personal relationships between the scientists and managers who were interviewed
Quorn	The trade name used by RHM to describe their texturised mycelial fungus grown on a glucose substrate, (mycoprotein)
Reference market	A market for an innovatory product based on an existing market (see Chapter 4)
Substrate	Substance upon which micro-organisms feed, breaking it down by enzyme action
Technology	Knowledge related to the use of physical objects or artefacts (see Chapter 10)

Notes

1 TECHNICAL INNOVATION AND THE ECONOMY

1 See Freeman *et al.* (1982) and Rothwell and Zegveld (1979) for a 'neo-Schumpeterian' explanation of observed changes in employment patterns in OECD countries since the war.
2 Although polymer science was developed after the discovery of the early synthetics, i.e. this is another case where the science came after the technology (Pavitt 1986).
3 See Chapter 7 conclusions for a discussion of this idea in relation to the findings of the research.
4 Here Teece draws on Polanyi's ideas of the nature of knowledge and technical knowledge in particular (Polanyi 1958).
5 Dosi has made a similar observation of 'selection rules' (Dosi 1982).
6 The economist Herbert Simon has perhaps made the strongest argument that the widespread use of what he calls 'rules of thumb' represents a challenge to simple assumptions about the rationality of decision-making (Simon 1959).

2 INTRODUCTION TO THE TECHNOLOGY

1 Also see Marstrand (1981) for an overview of SCP development.
2 Fungi are multicellular, but companies that fermented fungi attended the same conferences as those interested in yeast and bacterial fermentation. Their scientists also published papers in the same journals.
3 Much of this section is abstracted from ICI Educational Publications (1974).

3 THE MARKETS AND THE ENVIRONMENT DURING SCP DEVELOPMENT

1 Exactly the same kind of work went into developing 'Coffee Mate'.
2 See Chapter 7 for detail.
3 Food advisory groups.

4 COOPERATION AND COMPETITION BETWEEN COMPANIES

1 This was a GIE, Groupe d'Intérêt Economique, a specific form of corporate structure established in French (and since 1987, EEC) company law, e.g. Airbus Industrie is another.
2 United Nations Protein Advisory Group.

5 FIRM AND INDUSTRY CULTURE AND THE ROLE OF SENIOR MANAGEMENT

1 The name Finlay is used for the head of research.
2 A *Financial Times* science and technology journalist.
3 The Pruteen project champion, not the real name.
4 The real names of individuals have not been used in the text.
5 The Mezzogiorno is the southern, impoverished half of Italy.
6 Il Sole (1975) refers to the history of 'le bioproteine' as akin to a detective story in its complications and its secrecy.
7 BP had indeed had a temporary problem with emissions with its Lavera plant, but these had been solved. A BP manager working at an adjacent plant to the Lavera SCP plant told how there had been a flurry of press reports of cases of asthma allegedly linked to the SCP plant and its smell, which had been strong for a time.
8 Private conversation, 1988. David Sharp was gathering material for his forthcoming book on SCP.

6 TECHNICAL CHOICES IN THE DEVELOPMENT OF THE NOVEL FERMENTATION PROJECTS

1 The 1984 British Biotechnology Directory, which lists current biotechnology research in the UK, refers to the RHM work under the heading 'fungal single-cell protein'.
2 Seven years after their project began RHM changed from the penicillin to the A3/5 Fusarium strain with which they have worked ever since.
3 Although others, in SugarCo and ICI found this hard to believe.
4 See Bushell (1983) for a description of the Shell mixed culture.
5 In harvesting yeast and fungi, a BP microbiologist recalled that the main problem was educating engineers to design equipment such as the centrifuges for the minimum chance of infection of the biomass.
6 Studies of the creation of technology, or 'social construction' of technology, also highlight the inadequacy of the division between 'social' and 'technical', referring to the 'seamless web' character of society and technology (Bijker, Hughes and Pinch 1987: 9–10).

8 A SOCIO-COGNITIVE APPROACH TO INNOVATION

1 The term 'cognitive element' is used interchangeably with 'elements of understanding'.

APPENDIX

1 Although Senez, who had been invited to Japan several times to speak in favour of SCP, believed that the extraordinarily strong consumer lobbying against SCP in Japan was part supported by the American government, who had an interest in preserving their soya-meal export outlets.
2 The plant cost 80 million Swiss francs and took three to four years to design and construct.
3 At present they are ahead of requirements and have managed to cut the BOD of the effluent dumped in the River Aare by 96 per cent.

Bibliography

ACARD, (1982) *The Food Industry and Technology*, London: HMSO.

Adelman, C., Jenkins, D. and Kemmis, S. (1977) 'Rethinking Case Study: Notes from the Second Cambridge Conference', *Cambridge Journal of Education*, 6: 139–150.

Alford, B.W.E. (1976) 'The Chandler Thesis: Some General Observations', in Hannah, L. (ed.) *Management Strategy and Business Development*, London: Macmillan.

Allen, D.H. (1972) 'Credibility and the Assessment of R & D Projects', *Long Range Planning*, 5, 2: 53–64.

Allen, P.M. (1988) 'Evolution, Innovation and Economics', in Dosi, G., Freeman, CO., Nelson, R., Silverburg, T. (eds) *Technical Change and Economic Theory*, London: Pinter.

Allen, T.J. (1967) 'Communications in the R & D Laboratory', *Technology Review*, 70, 1: 31–37.

Antebi, E. and Fishlock, D. (1986) *Biotechnology: Strategies for Life*, Cambridge MA: MIT Press.

Arima, K. (1979) 'The Problems of Public Acceptance of SCP', in *Colloque International sur les Protéines d'Organismes Unicellulaire, Paris*, Paris: Technique et Documentation, pp. 145–162.

Atkinson, P. (1979) 'Ethnographic Research: An Outline,' in Open University (ed.) *Research Methods in the Social Sciences*, Milton Keynes: Open University Press.

Becker, H.S. and Geer, B. (1972) 'Participant Observation and Interviewing', in Manis, J.G. and Meltzer, B.N. (eds) *Symbolic Interaction*, Boston: Allyn & Bacon.

Beech, G.A. (1985) 'Food Drink and Biotechnology', in Higgins, I.J., Best, D.J. and Jones, J. (eds) *Biotechnology: Principles and Applications*, Oxford: Blackwell.

Behbehani, K., Handon, I.Y., Shams, A. and Hassin, W. (1978) 'Bioconversion Systems for Feed Production in Kuwait', in King, A. (ed.) *Bioresources for Development*, Oxford: Pergamon.

Bel Industrie (1986) *Bel Industrie Newsletter No. 7, June*, Paris: Bel Industrie.

Bessant, J. and Grunt, M. (1985) *Management and Manufacturing Innovation in the United Kingdom and West Germany*, Aldershot: Gower.

Bijker, W.E., Hughes, T.P. and Pinch, T.J. (eds) (1989) *The Social Construction of Technological Systems*, Cambridge MA: MIT Press.

Braun, E. (1981) 'Constellations for Manufacturing Innovation', *Omega*, 9, 3: 247–253.

Bull, A.T., Holt, G. and Lilly, M.D. (eds) (1982) *Biotechnology: International Trends and Perspectives*, Paris: OECD.

Burke, F. (1970) 'Logic and Variety in Innovation Processes', in Goldsmith, M. (ed.) *Technological Innovation and the Economy*, New York: Wiley.

Bushell, M.E. (1983) 'Application of the Principles of Industrial Microbiology to Biotechnology', in Wiseman, A. (ed.) *Principles of Biotechnology*, Brighton: Sussex University Press.

Cameron, S. (1980) 'New ICI "Bug" Boosts Plan for Protein Plant,' *Financial Times*, 11 April, p. 8.

Cartermill Publishing (1984) *The 1984 Directory of British Biotechnology*, St Andrews: Cartermill Publishing.

Cellulose Attisholz (1977) *Yeast Production by the Attisholz Fermentation System*, Attisholz, Solothurn, Switzerland: Cellulose Attisholz.

Centre for the Study of Industrial Innovation (1971) *On the Shelf: A Study of Industrial R & D Projects Abandoned for Non-Technical Reasons*, London: CSII.

Chandler, A.D. (1976) 'The Development of Modern Management Structure in the US and the UK', in Hannah, L. (ed.) *Management Strategy and Business Development*, London: Macmillan.

Chandler, A.F. (1962) *Strategy and Structure*, Cambridge, MA: MIT Press.

Chemical and Engineering News (1973) 'Health Issue Halts Petro-Protein Plans', *Chemical and Engineering News*. 5 March.

Clark, N. (1987) *The Political Economy of Science and Technology*, Oxford: Blackwell.

Cole, S. (1978) 'The Global Futures Debate', in Freeman, C. (ed.) *World Futures*, London: Heinemann.

Coombs, R. and Saviotti, P. (1987) *Economics and Technological Change*, London: Macmillan.

Cooper, A. and Schendel, D. (1976) 'Strategic Responses to Technological Threats', *Business Horizons*, February: 61–69.

Cronbach, L.J. (1975) 'Beyond the Two Disciplines of Scientific Psychology', *American Psychologist*, 30: 116–117.

Cyert, R.M. and March, J.G. (1963) *A Behavioural Theory of the Firm*, Englewood Cliffs: Prentice Hall International.

Daft, R.L. and Weick, K.E. (1984) 'Towards a Model of Organisations as Interpretation Systems', *Academy of Management Review*, 9, 2: 284–295.

Daly, P. (1985) *The Biotechnology Business: A Strategic Analysis*, London: Pinter.

Dansk Bioprotein (1988a), *Documentation: Bioprotein*, Odense: Dansk Bioprotein.

Dansk Bioprotein (1988b), *Introduction: Bioprotein*, Odense: Dansk Bioprotein.

Davy, C.A.E. and Wilson, D. (1980) 'Commercial Production of Feed Yeast from Carbohydrate Waste', in Moo-Young, M. (ed.) *Advances in Biotechnology*, Oxford: Pergamon.

Delbeke, J. (1984) 'Recent Long Wave Theories: A Critical Survey', in Freeman, C. (ed.) *Long Waves in the World Economy*, London: Pinter.

Dickson, K.E. and Puri, A. (1986) *Technical Change and International*

Competitiveness in the Food and Drink Industry, London: Technical Change Centre.

Dosi, G. (1982) 'Technological Paradigms and Technological Trajectories', *Research Policy*, 11: 147–162.

Duijn, J.S. van (1983) *The Long Wave in Economic Life*, London: Allen & Unwin.

Ebbinghaus, L., Ericsson, M. and Lindblom, M. (1980) 'Production of Single Cell Protein from Methanol by Bacteria', in Moo-young, M. (ed.) *Advances in Biotechnology*, Oxford: Pergamon.

Edelman, J., Fewell, A. and Solomons, G. (1983) 'Myco-protein: A New Food', *Nutrition Abstracts and Reviews*, 53, 6: 1–10.

Edelman, J. and Fewell, A. (1985) 'Commodities into Food', *Philosophic Transactions of the Royal Society of London B*, 310: 317–325.

Elger, A.J. (1975) 'Industrial Organisations: A Processual Perspective', in McKinlay, J.B. (ed.) *Processing People: Cases in Organisational Behaviour*, London: Holt, Rinehart & Winston.

Ericcson, M., Ebbinghaus, L. and Lindblom, M. (1981) 'Single-cell Protein from Methanol: Economic Aspects of the Norprotein Process', *Journal of Chemical Technology and Biotechnology*, 31: 33–43.

European Chemical News (1973) 'Petroprotein Plan in Japan – All Cancelled', *European Chemical News*, 2 March.

Fishlock, D. (1982) *The Business of Biotechnology*, London: Financial Times Business Information Service.

Foot-Whyte, W. (1960) 'Interviewing in Field Research', in Burgess, R.S. (ed.) *Field Research: A Source Book and Field Manual*, London: Allen & Unwin.

Freeman, C. (1982) *Economics of Industrial Innovation*, London: Pinter.

Freeman, C. (ed.) (1984) *The Long Wave in the World Economy*, London: Pinter.

Freeman, C., Clark, J.A. and Soete, L. (1982) *Unemployment and Technical Innovation: A Study of Long Waves and Economic Development*, London: Pinter.

Freeman, C. and Perez, C. (1986) Paper prepared for the Venice Conference, 1986, *The Diffusion of Technical Innovations and Changes of Techno-economic Paradigm*, mimeo: Science Policy Research Unit, University of Sussex.

Freeman, C. and Perez, C. (1988) 'Structural Crises of Adjustment, Business Cycles and Investment Behaviour', in Dosi, G., Freeman, C., Nelson, R., Silverburg, G. and Soete, L. (eds) *Technical Change and Economic Theory*, London: Pinter.

Gilfillan, S.C. (1963) *The Sociology of Invention*, Cambridge, MA: MIT Press.

Glaser, B.G. and Strauss, A.L. (1967) *The Discovery of Grounded Theory: Strategies for Qualitative Research*, Chicago: Aldine.

Gold, B. (1971) *Explorations in Management Economics*, London: Macmillan.

Gold, B. (1981) 'Technological Diffusion in Industry: Research Needs and Shortcomings', *The Journal of Industrial Economics*, 29, 3: 247–269.

Gold, B., Peirce, W.S., Rosegger, G. and Perlman, M. (1984) *Technological Progress and Industrial Leadership: The Growth of the US Steel Industry 1900–1970*, Toronto: Heath, Lexington Books.

Goldberg, (1985) *Single Cell Protein*, Berlin: Springer-Verlag.

Graham, M.B.W. (1986) *RCA and the Video Disc*, Cambridge: Cambridge University Press.

Greenshields, R. and Rothman, H. (1986) 'The Fermenter: Basic Concepts, Development and Types', in Jacobsson, S., Jamison, A. and Rothman, H. (eds) *The Biotechnological Challenge*, Cambridge: Cambridge University Press.

Griliches, Z. (1984) *R & D, Patents and Productivity*, Chicago: University of Chicago Press.

Hacking, A. (1986) *Economic Aspects of Biotechnology*, Cambridge: Cambridge University Press.

Halfpenny, P. (1979) 'The Analysis of Qualitative Data', *Sociological Review*, 27, 4: 799–825.

Hamer, G. (1979) 'Hydrocarbons in Biotechnology', in Harrison, D.E.F. (ed.) *Proceedings of a Meeting Organised by the Inst. Petroleum*, London: Heyden & Son.

Hamer, G. (1985) 'Chemical Engineering and Biotechnology', in Higgins, I.J., Best, D.J. and Jones, J. (eds) *Biotechnology: Principles and Applications*, Oxford: Blackwell.

Hamer, G., Harrison, D.E.F., Toplwald, H.H. and Gabriel, A. (1976) 'Natural Gas Processing and Utilisation', in The Institution of Chemical Engineering Symposium Series, *Natural Gas Processing and Utilisation*, 44: 65–72.

Hammersley, M. (1979a) 'Analysing Ethnographic Data', in Open University (ed.) *Research Methods in Education and the Social Sciences*, Milton Keynes: Open University Press.

Hammersley, M. (1979b) 'Data Collection in Ethnographic Research', in Open University (ed.) *Research Methods in Education and the Social Sciences*, Milton Keynes: Open University Press.

Hannah, L. (1976) *The Rise of the Corporate Economy*, London: Methuen.

Hannah, L. (1984) 'Entrepreneurs and the Social Sciences', *Economica*, 51: 219–234.

Harris, R. (1987) *Power and Powerlessness in Industry: An Analysis of the Social Relations of Production*, London and New York: Tavistock.

Harvey-Jones, J. (1988) *Making it Happen: Reflections on Leadership*, London: Collins.

Hayes, R.H. and Abernathy, W.J. (1982) 'Managing Our Way to Economic Decline', in Tushman, M.L. and Moore, W.L. (eds) *Readings in the Management of Innovation*, London: Pitman.

Hesseltine, C.W. and Wang, H.L. (1980) 'Fermented Foods', *Food Trade Review*, September.

Hickson, D.J., Butler, R.J., Cray, D., Mallory, G.R. and Wilson, D.C. (1986) *Top Decisions: Strategic Decision-Making in Organisations*, San Francisco: Jossey-Bass.

Higgins, I.J., Best, D.J. and Jones, J. (eds) (1985) *Biotechnology: Principles and Applications*, Oxford: Blackwell.

Holmes, A.W. and Jarvis, B. (1981) *The Application of Biotechnology to the Food Industry*, Leatherhead: Leatherhead Food Research Association.

Howells, E. (1982) 'Opportunities in Biotechnology for the Chemical Industry: Single Cell Protein and Related Technology', *Chemistry and Industry*, 7 August.

ICI (1974) *Pruteen: A New Protein Source*, Billingham: Billingham Press.

ICI (1984) *Pruteen*, Billingham: ICI.

ICI Educational Publications (1974) *New Protein*, Birmingham: The Kynoch Press.

Il Sole (1975) 'Nuove Indagini per le Bioproteine,' *Il Sole*, 20 June, p. 3.

Jasanoff, S. (1985) 'Technological Innovation in a Corporatist State: The Case of Biotechnology in the Federal Republic of Germany', *Research Policy* 4: 23–38.

Jenkins, G. (1985) 'The BP Protein Process: A Case Study', *International Industrial Biotechnology*, 6: 6. 1–4.

Johnson, M.J. (1972) 'Technology for Selection and Evaluation of Cultures for Biomass Production,' *Fermentation Technology Today*: 473–477.

Kane F. (1993) 'Hunt is on for Quorn in Sausage and Meat Pie Country', *Guardian*, 15 April, p. 3.

Kay, N. (1979a) 'Corporate Decision-Making for Allocations to Research and Development', *Research Policy*, 8: 46–49.

Kay, N. (1979b) *The Innovating Firm*, London: Macmillan.

Kay, N. (1988) 'The R & D Function: Corporate Strategy and Structure', in Dosi, G., Freeman, C., Nelson, R., Silverburg, G. and Soete, L. (eds) *Technical Change and Economic Theory*, London: Pinter.

Kennedy, C. (1986) *ICI: The Company that Changed Our Lives*, London: Hutchinson.

King, P.P. (1982) 'Biotechnology: An Industrial View', *Journal of Chemical Technology and Biotechnology*, 32: 2–8.

Knorr, D. and Sinskey, A.J. (1986) 'Biotechnology in Food Production and Processing', in Koshland, D. (ed.) *Biotechnology, the Renewable Frontier*, Washington, D.C.: American Association for the Advancement of Science.

Komsomolskaya Pravda (1988) 'Bacterial Paprine Bomb', 15 March, p. 2.

Kotter, J.P. (1982) *The General Managers*, New York: The Free Press.

Kuhn, T.S. (1970) *The Structure of Scientific Revolutions*, Chicago: University of Chicago Press.

Laine, B.M., Snell, R.C. and Peet, W.A. (1976) 'Production of Single Cell Protein from n-Paraffins', *The Chemical Engineer*, June: 440–4.

Landau, R. and Rosenberg, N. (1986) 'Editors' Overview', in Landau, R. and Rosenberg, N. (eds) *The Positive Sum Strategy: Harnessing Technology for Economic Growth*, National Academy Press.

Langrish, J., Gibbons, M. Evans, W.G. and Jevons, F.R. (1972), *Wealth from Knowledge: A Study of Innovation in Industry*, London: Macmillan.

Litchfield, J.H. (1983), 'Single Cell Proteins,' *Science*, 219: 740–746.

Lothian, N. (1984) *How Companies Manage R & D*, London: Institute of Chartered Management Accountants.

Lunn, J. and Martin, S. (1986) 'Market Structure, Firm Structure and Research and Development', *Quarterly Review of Economics and Business*, 1, 26: 31–34.

Macdonald, J.M. (1985) 'R & D and the Directions of Diversification', *Review of Economics and Statistics*, 4, 67: 583–590.

Mangham, I. (1986) Power and Performance in Organisations, Oxford: Blackwell.

Mansfield, E. (1977) *The Production and Application of New Industrial Technology*, New York: Norton.

Mansfield, E. (1981) 'How Economists See R & D,' *Harvard Business Review*, November/December: 23–29.

Marlow Foods (1987) *Quorn: A Nutritional Guide*, Banbury: Marlow Foods.

Marstrand, P.K. (1981) 'Production of Microbial Protein: A Study of the Development and Application of a New Technology', *Research Policy*, 10, 2: 148–171.

Marstrand, P.K. and Rush, H.J. (1979) 'Shadows on the Seventies: Indicative World Plan, the Protein Gap and the Green Revolution', in Whiston, T. (ed.) *The Uses and Abuses of Forecasting*, London: Macmillan.

Medawar, P. (1979) *Advice to a Young Scientist*, London: Harper & Row.

Metcalfe, S. (1988) 'The Diffusion of Innovation: An Interpretative Study', in Dosi, G., Freeman, C., Nelson, R., Silverburg, G. and Soete, L. (eds) *Technical Change and the Economy*, London: Pinter.

Mintzberg, H. (1973a) 'An Emerging Strategy of "Direct" Research', in van Maanen, (ed.) *Qualitative Research*, Beverley Hills: Sage Publications.

Mintzberg, H. (1973b) *The Nature of Managerial Work*, New York: Harper & Row.

Mintzberg, H. and Waters, J.A. (1985) 'Of Strategies, Deliberate and Emergent', *Strategic Management Journal*, 6: 257–272.

Mishler, E.G. (1986) *Research Interviewing, Context and Narrative*, Cambridge, MA: Harvard University Press.

Mitchell, J.C. (1983) 'Case and Situation Analysis', in *Sociological Review*, 31, 2: 187–211.

Moo-Young, M. (1978) 'A New Source of Feed and Food Proteins: The Waterloo SCP Process', in King, A. (ed.) *Bioresources for Development*, Oxford: Pergamon.

Moo-Young M. (1980), 'Advances in Biotechnology,' *Proceedings of the 6th International Fermentation Symposium*, vol. 2, Oxford: Pergamon.

Moses, V. and Rabin, B. (1982) 'EIV Special Report Number 124', *Biotechnology: A Guide for Investors*: 68–70.

Mowery, D.C. and Rosenberg, N. (1979) 'The Influence of Market Demand on Innovation: A Critical Study of Some Recent Empirical Studies', *Research Policy*, 8: 103–153.

Mulkay, M.J. (1972) *The Social Process of Innovation: A Study in the Sociology of Science*, London: Macmillan.

Naslund, B. and Sellstedt, B. (1974) 'Budgets for Research and Development: An Empirical Study of 69 Swedish Firms', *R & D Management*, 4, 2: 67–73.

Neale, W.C. (1984) 'Technology as Social Process: A Commentary on Knowledge and Human Capital', *Journal of Economic Issues*, 18, 2: 573–580.

Nelson, R.R. (1968) 'Innovation', in Sills, D.S. (ed.) *International Encyclopaedia of the Social Sciences*, London: Collier Macmillan.

Nelson, R.R. (1986) 'Institutions Supporting Technical Advance in Industry', *American Economic Review*, 76: 186–189.

Nelson, R.R. and Winter, S.G. (1977) 'In Search of a Useful Theory of Innovation', *Research Policy*, 6: 36–76.

Norris, K. and Vaizey, J. (1973) *The Economics of Research and Technology*, London: Allen & Unwin.

Norris, K.P. (1971) 'The Accuracy of Project Cost and Duration Estimates in Industrial R & D', *R & D Management*, 2, 1: 25–36.

Olson, S. (1987) *Biotechnology: An Industry Comes of Age*, New York: National Academy Press.

Pavitt, K. (1980) *Technical Innovation and British Economic Performance*, London: Macmillan.

Pavitt, K. (1981) 'Technological Innovation in the UK: A Suitable Case for Improvement,' in Carter, C. (ed.) *Industrial Policy and Innovation*, London: Heinemann.

Pavitt, K. (1986) 'Technology, Innovation and Strategic Management', in McGee, J. and Thomas, H. (eds) *Strategic Management Research*, New York: Wiley.

Peter, M.F. (1983) 'The Technological Dimension of Competitive Strategy', *Research on Technological Innovation Management and Policy*, 1: 1–33.

Pettigrew, A. (1977) 'Strategy Formulation as Political Process', *International Studies of Management and Organisation*, 7, 2: 78–87.

Pettigrew, A. (1979) 'On Studying Organisational Cultures', in van Maanen J. (ed.) *Qualitative Methodology*, Beverly Hills: Sage Publications.

Pettigrew, A. (1985) *The Awakening Giant: Continuity and Change in ICI*, Oxford: Blackwell.

Piore, M.J. (1979) 'Qualitative Research Techniques in Economics', in van Maanen, J. (ed.) *Qualitative Methodology*, Beverly Hills: Sage Publications.

Polanyi, M. (1958) *Personal Knowledge: Towards a Post Critical Philosophy*, London: Routledge & Kegan Paul.

Popper, K. (1972) *The Logic of Scientific Discovery*, London: Hutchinson.

Porter, M.E. (1983) 'The Technological Dimension of Competitive Strategy', in *Research on Technological Innovation, Management and Policy*, 1: 1–33.

Prentis, S. (1984) *Biotechnology: A New Industrial Revolution*, Maryknoll, NY: Orbis.

Quinn, J.B. (1988) 'Managing Strategies Incrementally', in Quinn, J.B. and Mintzberg, H. (eds) *The Strategy Process*, Englewood Cliffs, NJ: Prentice Hall.

Radnor, M. and Rich, R. (1980) 'Organisational Aspects of R & D Management: A Goal-Directed Contextual Perspective', in Dean, B.V. and Goldhar, J.L. (eds) *Management of Research and Innovation*, Amsterdam: North Holland.

Rimmington, A. (1984) 'Biotechnology in the USSR', *Industrial Biotechnology*, 3, 8: 1–15.

Rimmington, A. (1985a) 'Single Cell Protein: The Soviet Revolution', *New Scientist*, 27 June, pp. 12–15.

Rimmington, A. (1985b), 'Biotechnology in the USSR – A Reassessment', *Industrial Biotechnology*, 4: 33–39.

Riviere, J. (1977) *Industrial Applications of Microbiology*, Guildford: Surrey University Press.

Roberts, E.B. (1968) 'The Myths of Research Management', *Science and Technology*, August: 40–46.

Romantshuk, H. (1975) 'The Pekilo Process: Protein from Spent Sulphite Liquor', in Tannenbaum, S.R. and Wang, D.I.C. (eds) *SCP 2*, Cambridge, MA: MIT Press.

Romantshuk, H. and Lehtomaki, M. (1978) 'Operational Experiences of First Full Scale Pekilo SCP Mill Application', *Process Biochemistry*, 29: 16–17.

Rosenberg, N. (1976) *Perspectives on Technology*, Cambridge: Cambridge University Press.

Rosenbloom, R.S. and Abernathy, W.J. (1982) 'The Climate for Innovation in Industry', *Research Policy*, 11: 209–225.

Rothwell, R. (1977) 'The Characteristics of Successful Innovators and Technically Progressive Firms', *R & D Management*, 7, 3: 191–206.

Rothwell, R. and Zegveld, W. (1979) *Technical Change and Employment*, London: Pinter.

Routh, G. (1984) *Economics: An Alternative Text*, London: Macmillan.

Rubinstein, A.H. and Chakrabati, A.K. (1976) 'Factors Influencing Innovation Success at the Project Level', *Research Management*, 19: 15–20.

Scherer, F.M. (1984) *Innovation and Growth: Schumpeterian Perspectives*, Cambridge, MA: MIT Press.

Schmookler, J. (1962) 'Changes in Industry and the State of Knowledge as Determinants of Industrial Invention', in NBER (ed.) *The Rate and Direction of Inventive Activity*, Princeton: Princeton University Press.

Schott, B. (1974) 'R & D, Innovation and Microeconomic Growth: A Case Study', *Research Policy*, 2: 380–403.

Schumpeter, J.A. (1942) *Capitalism, Socialism and Democracy*, New York: Harper & Row.

Science Policy Research Unit (1972) *Success and Failure in Industrial Innovation*, London: Centre for the Study of Industrial Innovation.

Sharp, M. (1985a) *The New Biotechnology*, Sussex: Sussex European Papers No. 15.

Sharp, M. (1985b) 'Biotechnology: Watching and Waiting', in Sharp, M. (ed.) *Europe and the New Technologies*, London: Pinter.

Sherwood, M. (1974) 'Single Cell Protein Comes of Age', *New Scientist*, 28 November, pp. 634–639.

Silver, S. (1986) *Biotechnology: Potentials and Limitations*, Berlin: Springer.

Simon, H.A. (1955) 'A Behavioural Model of Rational Choice', *Quarterly Journal of Economics*, 69: 99–118.

Simon, H.A. (1959) 'Theories of Decision Making in Economics and Behavioural Science', *American Economic Review*, 49: 253–283.

Smircich, L. (1985) 'Is the Concept of Culture a Paradigm for Understanding Organisations and Ourselves?' in Frost, P.S., Moore, L.F., Louis, M.R., Lundberg, C.C. and Martin, J. (eds) *Organisational Culture*, London: Sage.

Smith, S.R.L. (1980) 'Single Cell Protein', *Phil Trans Royal Society of London* B, 290: 341–354.

Solomons, G.L. (1983) 'Single Cell Protein', *Critical Reviews in Biotechnology*, 1: 21–58.

Spencer, C. (1989) 'Turning a New Leaf', *Guardian*, 5 August, p. 15.

Spicer, A. (1971) 'Synthetic Proteins for Human and Animal Consumption', *The Veterinary Record*, 89: 482–486.

Spinks, A. (1982) 'Targets in Biotechnology', *Proceedings of the Royal Society of London* B, 214: 289–303.

Spradley, J.P. (1979) *The Ethnographic Interview*, London: Holt, Rinehart & Winston.

Stankiewicz, R. (1981) *The Single Cell Protein as a Technological Field*, Lund: AV Centralem.

Stanley, R. (1977) *Single Cell Protein: An Overview*, Mimeo, University of Aston Technology Policy Unit.

Steinkraus, K. (1980) 'Production of Microbial Protein Foods on Edible Substrates, Food By-products and Lignocellulosic Waste', in King, A. (ed.) *Bioresources for Development*, Oxford: Pergamon Press.

Strauss, A.L. (1987) *Qualitative Analysis for the Social Sciences*, Cambridge: Cambridge University Press.

Teece, D.J. (1982) 'Towards an Economic Theory of the Multiproduct Firm', *Journal of Economic Behaviour and Organisation*, 3: 39–63.

Teece, D.J. (1988) 'Technological Change and the Nature of the Firm', in Dosi, G., Freeman, C., Nelson, R., Silverburg. G. and Soete, L. (eds) *Technical Change and Economic Theory*, London: Pinter.

Thomas, H. (1970) 'Economic and Decision Analysis: Studies in R & D in the Electronics Industry', unpublished PhD thesis, University of Edinburgh.

Turner, B.A. (1981) 'Some Practical Aspects of Qualitative Data Analysis: One Way of Organising the Cognitive Processes Associated with the Generation of Grounded Theory', *Quality and Quantity*, 15: 225–247.

Utterback, J.M. (1974) 'The Dynamics of Product and Process Innovation', in Hill, C.T. and Utterback, J.M. (eds) *Technological Innovation for a Dynamic Economy*, Oxford: Pergamon Press.

Walker, M. (1988) 'Bio-plant Poisons Soviet Town', *Guardian*, 16 March.

Watson, T.J. (1986) *Management Organisation and Employment Strategy*, London and New York: Routledge, Kegan & Paul.

Wegner, G.H. (1983) 'Nonpolluting Process Turns Methanol into Protein', *Chemical Engineering*, 5 September, pp. 56–57.

Weick, K.E. (1979) *The Social Psychology of Organising*, Reading, MA: Addison-Wesley.

Weick, K.E. (1985) 'The Significance of Corporate Culture,' in Frost, P. (ed.) *Organisational Culture*, London: Sage.

Wells, J. (1975) 'Analysis of Potential Markets for Single Cell Protein,' in *Symposium on Single Cell Protein*, Philadelphia, PA: American Chemical Society, 9 April.

Wiles, P. (1984) 'Epilogue: The Role of Theory', in Wiles, P. and Routh, G. (eds) *Economics in Disarray*, Oxford: Blackwell.

Williams, R. (1984) 'The International Political Economy of Technology', in Strange, S. (ed.) *Paths to International Political Economy*, London: Allen & Unwin.

Wilson, M.J. and Bynner, J. (1979) 'Variety in Social Science Research', in Open University (ed.) *Research Methods in Education and the Social Sciences*, Milton Keynes: Open University.

Winkofsky, E.P. and Mason, M. (1980) 'R & D Budgeting and Project Selection: A Review of Practices and Models', in Dean, R.V. and Goldhar, J.L. (eds) *Management of Research and Innovation*, Amsterdam: North Holland.

Wright Mills, C. (1970) *The Sociological Imagination*, Harmondsworth: Penguin.

Index

Aare, River 215n
Abernathy, W.J. 13–14
acid rain 200
acids 29, 138, 142; acetic 198, 199;
    citric 102, 104, 131; fatty 102, 124;
    formic 141; lactic 205; phosphoric
    25; propionic 124–5; sulphuric 25,
    197; uric 24, 154; *see also* amino
    acids; nucleic acids; pH values
Adelman, C. 10, 12
additives 19, 58, 62
aeration 58, 205
aflatoxins 150
agar 23, 25, 27
agriculture 21, 131, 207
air 143, 145; recycling of explosive
    methane–air mixtures 142;
    oxygen-depleted 29
Alaska 115
alcohol 26, 36, 89, 199, 205
aldehydes 142
algal 136
Algeria 66, 67
alkali 151
alkanes 54, 65, 72, 153; fermentation
    of 124; gaseous 142; undigested
    substrate 144; yeast grown on 80;
    see also n-alkanes
amino acids 31, 32–3, 45, 46, 76, 135;
    essential, sulphur-containing 41;
    see also lysine; methionine
ammonia 84, 111, 131, 148; aqueous
    24–5
Amsterdam 96, 97
d'Angelo, Tony 61

ANIC (Italian chemicals combine)
    22, 116, 156
animal feed 17, 20–1, 98, 150, 188;
    additive 19; bulk 41, 99;
    characteristics affect the design of
    the product 46–7; citrus waste into
    209; consumption requirements
    193; contaminated with polycyclic
    hydrocarbons 188; demand for 54;
    grain for 66, 192; high-protein 18,
    193, 208; lack of protein for 193;
    microbial feeds as 69; novel 69,
    151; poor quality 193; positive
    economic value as 197; post-war
    shortage 202; prices 41, 53;
    proteins 18, 148, 193, 208; SCP for
    40–1, 99, 186; second-largest
    company in the world 126;
    skimmed-milk powder 186; *see
    also* animal feed market; feed
    compounders; fish-meal; soya-
    meal
animal feed market 31, 37, 51, 56–9,
    110, 205; domestic 206; high-
    value-added 187; new idea of, the
    guide for research 49; piglets 56,
    186; poultry 56; production of SCP
    for 40–1; SCP companies all
    attempting to obtain share 71;
    semi-moist 118; shock loss of 58–9,
    206; understanding 45; western
    41–4
Antebi, E. 132
antibiotic business 161
aquaculture 59, 206

Argentina 41, 55
Arima, K. 195, 196n
aromatic compounds 124
asepsis 138–40, 143, 144, 162
ash 133
Aspergillus Niger 135, 161–2, 208, 209
Association des Souches 205

backhanders 119
bacon rashers 42
bacteria 18, 25, 48, 84, 135, 142; absolutely stable 140; biochemical pathways 24; difficult to centrifuge 27; harvesting from solution involved a flocculation step 29–31; hydrogen-utilising 158; infection 138, 139, 150; methane-utilising 94, 141; more efficient 159; optimal growth of 47; protein content 27, 28, 31, 58, 148; reproduction 23; speed of growth 27
bacteriology 141
bagasse 89, 91, 130, 192
bakeries 207
Bassetts 50
'BBK' 190
beef 60, 61
Bel Industrie 43, 44, 58–9, 204–7
Belgium 3
Belize 90–1, 135, 209
benzene 26, 102
Berne 200
Bessant, J. 15
Bijker, W.E. 214n
Billingham *see* ICI Biological Products Division
biochemical pathways 24, 75, 124, 137, 158, 159
biochemistry 89, 171
biodegradable products 94, 95, 118, 158
biological sciences 85, 97; research 111, 112, 171
biomass 130, 136; alternative uses for 161; doubts on non-sterile fermentation 140; harvesting 147–9; production 36, 140; texturising 161; yeast 19, 59, 197

biopolymers 97, 156
bioprotein manufacture 20
biosurfactants 156
biosynthesis 131
biotechnology 21, 140, 156, 157, 171; development of 155; early 92; industrial 132, 149, 192; projects 85; public–private undertakings 99; research in 97; skill base 159; strategic interest in 95
Birmingham University 188
Biscuit, Cake, Chocolate and Confectionery Alliance 44
BMA (British Medical Association) 60
BOD (biological oxygen demand) 129, 197, 198–9, 202, 203, 215n
Bologna 121
Boregaard 199, 200
bounded rationality 15
BP (British Petroleum) 21, 23, 29, 55–6, 67, 72–6 *passim*, 87, 150, 155, 161, 184, 191, 195; alkane process 153; choice of organism influencing 45; costly Italian disaster 64, 156, 194; dearth of possible projects that linked core business to microbiology 157; determination and success in developing technology 130–1; engineering to sterile standards 139; fermentation achievements 143–4; gas–oil process 138; initial attraction of SCP 114; media distrust of toxicological claims 103; micro-organism investigation 25; new idea of animal feed market the guide for research 49; new regulatory guidelines for SCP 79; oil crises and 69; public perception of protein gap 64–5; screening of organisms 135; second-largest animal feed company in world 126; technology 196; testing in independent institutes 152; value of R&D 127; yeasts 80, 125, 159, 168, 193; *see also* ANIC; Champagnat; Italproteine; Toprina; *also BP divisions below*

BP Biological Sciences group 156, 157
BP France (Lavera) 44, 65, 115, 123, 140; contract for microbial cleaning system 37; gas–oil process 153, 187; SCP product approved by French government 154; temporary problem with emissions 214n; use of non-sterile mixed culture 141
BP Grangemouth 'semi-industrial' plant 37, 115, 116, 118, 156
BP International 114, 126
BP Nutrition 126, 157
BP Oil 116
BP Proteins 115, 120, 126
BP Sarroch (Sardinia) 64, 115–23 *passim*, 139, 141, 144, 156, 157; construction 65; closure 79, 99; plant dismantled 37
BP Sunbury Research Centre 114, 156
bran fibre 62
Braun, E. 15
Brazil 36, 41, 55
breweries 59, 206, 207
bribes 120
British Biotechnology Directory (1984) 214n
British Gas 104
British Technology Group 77
broiler feeds 41
bronchial asthma 189
Bull, A.T. 159
bureaucracy 120
burgers 63, 86, 161
Burke, F. 10
Bushell, M.E. 214n
by-products 196
Byelorussian republic 190

Cadbury-Schweppes 62
Cairo University 160
Camembert 141
Candida: Albicans 24, 103, 125; *Lippolytica* 24, 125; *Maltosa* 125; *Tropicalis* 24, 76, 103; *Utilis* 18, 19, 202
capital: costs 34–5, 36–7, 106, 117, 137, 188, 192, 199; expenditure 199; investment 137, 187, 204, 205; large returns on 116; unaided, private 207–8
carbohydrates 25, 26, 207; cellulose 197; waste 18, 20, 50, 196, 197, 199
carbon 26; conversion 27, 28, 139, 145; natural sources 160
carbon dioxide 24, 136, 148, 197; carbon substrate completely oxidised to 28; waste 29
carcinogens 118, 154, 190
carob projects 49, 90, 170, 173, 208
cascade theory 61
casein 58
catalysts 73, 84, 111, 156, 161
cattle feeds 41
cause maps 8, 9, 10, 163, 165, 175; company-wide 177
cells: bacterial, smaller 148; dead 28; density 135–6, 162; granules of PHB in, as means of storing energy 158; growth 24; mass 23, 133, 136, 154; separating from fermenter liquor 147; temperature and pH shock to 148; yeast 124, 126, 148
Cellulose Attisholz 23, 174, 200, 201n, 202, 203, 204
centrifuges 31, 149, 191, 148, 185
cereals 55
CFR 73
Chaim, Ernest 141
Champagnat (BP research scientist) 19, 37, 49, 115, 168–9
champions: product 49; project 60, 166–79, 214n
Chandler, A.F. 9
charcuterie 59, 207
Chartres 59
cheese 28, 62, 141
Chemical and Engineering News 194
chemical engineering 4, 84–6, 157
Chernobyl 189
chickens 42, 48, 60, 61, 152, 209; feed market 43, 46; liver necrosis in 154
citric acid 102, 104, 131
Club of Rome 53, 55
cobalt 47
'Coffee Mate' 213n

cognitive elements 175–6, 182, 183–4, 185

cognitive ensembles 183, 184, 185

Cole, S. 53

collaboration 78–80, 100

colour 72

Committee on Novel and Irradiated Foods 151

companies 22–3; cooperation and competition between 71–81, 214n; relationship between R&D and 127–8; structure 88; *see also* firms; *also under individual company names, e.g.* BP; ICI; RHM; Shell

competition 2, 71–81, 103, 181, 207, 214n; biotechnological 99

competitive advantage 207

confectionery 20, 196

conservatism 61, 170

constellation theory 15

consumers 28, 77, 194

contamination 27–8, 118–19, 188, 205

continuous fermentation 21, 141, 142–3, 162, 205; large-scale 134; monoseptic 76–7, 144, 145; switch from batch to 38

cooling equipment 137

cooperation 71–81, 214n

copper 150

core business 5, 6, 84, 96, 157, 162

corn cobs 192

corrupt connivance 122

costs 145, 147; capital 34–5, 36–7, 106, 117, 137, 188, 192, 199; development 199; energy 199; feedstock 139; hydrocarbon 66; low, alternative source of protein 100; marginal 43; operating 137, 199; pilot plant 99; plant construction 139; plant unable to cover 115; research 73; sodium and potassium hydroxide 151; transport 66

cotton 192, 193

Cowen 210

crop sprays 55

crude oil 26, 37, 116, 124, 169; first fractionation 114; price of 36, 52; truly mixed substrate of 154; with

a very high n-alkane content 114

culture 9; firm and industry 82–128, 181, 214n

cultures 76, 77, 138; bacterial, problem of infection when working with 139; contaminated with a foreign yeast strain 205; continuous 143; experimental, growing 22–3; high productivity 135; mixed and pure 140–2, 162

Cyert, R.M. 7

Cyprus 90, 208

Dainippon Ink and Chemicals Inc 118, 193, 194, 195, 196

dairy-free chocolate 59, 206

dairy producers 20

Dalgety 41

Daly, P. 132

Dansk Bioprotein 20, 80, 131–2, 136, 149, 185, 186–8; comany's approach to regulatory authorities 153; methane-based projects 139, 141; mixed culture 142; optimum growth temperature of organism 137; original business plan 43; stir-tank fermenters 144

degenerative allergies 189

Delbeke, J. 1

demand 54, 55, 149, 175

Denmark *see* Dansk Bioprotein

Department of Health 48

depreciation 106

desulphurisation 204

dewatering costs 135, 136

diesel fuel 114

diets 59, 60, 151, 208; high percentage SCP, response of animals to 154; part-Pruteen 42; shortage of protein in 20

disaster anecdotes 66

disease 28; immunity to 189

diversification 37, 96, 118; through R&D 87–90, 94

DNA (dioxyribonucleic acid) 23, 24, 154

dog coats 59, 206

Don, River 50

Dosi, G. 213n

drying processes 31, 135, 149, 205

Du Pont 77, 86–7, 100
Duijn, J.S. van 1

ease of use 183
Ebbinghaus, L. 35n
EC (European Community) 3, 115, 214n; dairy quotas 44; dried skimed milk 43, 106, 110, 186; food safety authorities 79; SCP testing procedures 80, 150
economics *see* capital; costs; investment; prices
economies of scale 4, 75, 109, 147, 210
Edelman, Jack 99
effluent 158, 198, 202, 203; sucrose-rich 50
eggs 62
electric current 31
electrodes 155
ELF 73
Elger, A.J. 8, 176
emulsion 24, 26
energy 20, 46, 159, 199, 205; chemical source 131
ENI Chemicals 39, 104, 116
entrepreneurs 2, 177–8, 179
enzymes 137, 148, 156, 160
enzymology 171
Ericcson, M. 35n
Escherichia Coli 159
Esso 73
ethane 26
ethanol 157, 192, 198, 199, 200, 204; *Saccharomyces* 202
ethene 175
ethics 119, 151
European Chemical News 194
EXCO 88, 89, 93
exhaust gas flues 189
extra-cellular products 24, 28

fat 61, 62, 63
fatty acids 102, 124
feed compounders 34, 41–2, 53, 69, 210
fermentation 47, 50, 75, 89, 92, 97; alcohol 36, 199; alkane 124, 143; anaerobic 198; aseptic 138–40, 143, 144, 162; bacterial 148; batch 142; carbohydrate wastes 196; carob 90, 208; choices 142–7; controlled 197; ethanol 200; large-scale 23, 134; liquid 135; local 207; low-technology 209; massive-scale 144; means of producing enzymes 160; methane 172; mixed culture 140–2; molasses 209; monoseptic 76–7, 144, 145; mycoprotein 38; necessary to remove sugars from liquor 202; non-sterile biomass 140; novel technology 129; paraffins 191; pilot 101, 134; protein 26; pure culture 140–2, 162; SCP plants 117; single-celled organisms on hydrocarbon substrates 19; small-scale 134; solid-state 135; starch-fungus 38; sterile 28, 148; substrates 20, 90; technical choices in development of novel projects 129–65, 214n; testing different substrates for suitability 90; two-stage 198; uncontaminated 144; using polymers for tertiary oil recovery by 131; using solvents 188; wish to develop sophisticated technology 131; wood pulp waste 197–204; yeast 199; *see also* continuous fermentation; fermentation technology; fermenters
fermentation technology 17, 68, 72, 126, 175, 192; key part of developing cluster of biotechnologies 175; novel 129; sophisticated 131, 144; starch-fungus 38
fermenters 46–7, 145–7, 149, 159; air-lift 29, 144–5; concrete hole-in-the-ground 51, 68; non-sterile 160; pilot plant 27; pressure cycle 29, 144–5; scaling-up 155; separating cells from liquor 147; stir-tank 144; tower 161; very simple 49
fertilisers 105, 207
fertility 189
fibre 62
filtration 27, 135; ultra- 205
Financial Times 187, 214n

Finland 200; environmental
  legislation 198; pulp industry 197
firms: characterised by technical
  knowledge 184; competition 2;
  culture and senior management
  82–128, 181, 214n; differences in
  spending between 6; environment
  9; innovating 5–6; knowledge base
  5; meaning of changes in
  environment 8; production
  technology 163–6, 182, 184, 191;
  socio-cognitive view of 182–4;
  technical change within 15
First World War 18
fish 62; aquaculture of 206; diets 152;
  fed on a part-Pruteen diet 42;
  market for feed 206; oil 22; price
  of 54
fish-meal 32, 41–3 *passim*, 47–8, 69,
  71, 106, 150; cheap 200; price of
  34, 186; protein product
  progressively substituted for 32
Fishlock, D. 34n, 95, 126, 132
flash-drying 31
flocculation 29–31
foaming agents 161
fodder: cattle 205; yeast 33, 58, 198,
  199, 202; *see also* animal feed
food 17, 61–2, 85–6; casein in 58;
  conservative concepts 62; ethical
  company 151; flavouring 57; high-
  protein 18, 63; low-fat and low-salt
  63; peripheral market 61;
  problems of having Third World
  populations accept a new source of
  210; reputation for 83–4; safety
  authorities 79, 149; techniques
  specific to the industry 160; world
  economic price relation between
  hydrocarbons and 67; *see also*
  animal feed; diets; human food;
  novel foods; supplements;
  vitamins
Foot-Whyte, W. 16
formaldehyde 141
fractionation 38, 58, 102, 110, 114
France 44, 49, 120, 206;
  development of biotechnology
  155; French Yeast Association
  205; Groupe d'Intérêt

Economique 214n; Groupe en
  France de Protéines 73;
  Kondratiev waves and technical
  innovation 3; Laboratoire
  National Agronomique de Paris
  205; Ministry of Agriculture 115;
  skimmed milk replacer market 44;
  see also BP France; IFP
Freeman, C. 2, 3n, 4, 5n, 184, 213n
Freudian theory 12
fungi 18, 25, 27, 45, 64, 132, 138;
  amino acid profiles 135; can grow
  on a range of substrates 135;
  filamentous 45, 75, 135; mycelium
  21; percentage of protein in 28;
  processes 147–8; single cell 23;
  starch-, fermentation technology
  38
Fusarium 38, 76, 214n

galactose 207
gas–oil 25, 26, 102, 115, 143, 187–8
gelling agents 161
genes 159
genetic engineering 140, 158
Germany 18, 59, 200, 206; animal
  feed produced 98; Kondratiev
  waves and technical innovation 3;
  Ministry of Foreign Development
  160; public–private undertakings
  in biotechnology 99; SCP-
  manufacturing technology 19;
  synthetic protein 98; *see also*
  Hoechst
Gildenburg 98, 99
Giste-Brocades 96–7
Glaser, B.G. 16
glasnost 189
Glaxo 92, 173
glucose 20, 21, 26
Gold, B. 7, 11, 12–13
Goldberg 158
goldfish 41
Gorki 188
gout 24
Graham, Margaret 13
Grangemouth *see* BP Grangemouth
granules 31, 46–7, 158
Greece 90
Greenshields, R. 146

Grocer 61
grounded theory 16
Grunt, M. 15
Guardian, The 189

Hacking, A. 34, 35n, 36n, 50
Hamer, G. 41, 85, 94, 133, 138
Harvey-Jones, Sir John 108, 191
health 59, 60; significant threat to 194
heat pumps 137
heavy metals 47, 48
Helsinki 197
heuristics 7, 10, 162
Hickson, D.J. 16
High Wycombe 78, 101
Hoechst 45, 106, 110, 141, 152, 160; biotechnology department 161; government technology strategy 98–9; Schlingmann process 148–9, 154–5; screening of organisms 135
Holt, G. 159
Hughes, T.P. 214n
human food 38, 45, 103, 122, 150; additives 58; desirable qualities 57; markets 58–9, 73, 99, 204; prospect of SCP as 154; protein and protein, useless as 148; supplements 87, 190, 205; yeast protein as 202, 204
hydrocarbons 18, 20, 33, 41, 59, 137; aromatic 26, 124; costs 66; major new technology based on 73; markets 53; polycyclic, feed contaminated with 188; prices 20, 51–3, 113, 184, 185; projects 23, 95; substrates 19, 20, 194; *see also* alkanes; gas–oil
hydrogen 26, 158
hydrogen sulphide 197
hydrolysed wood chips 192
hydrostatic pressure 29

ICI (Imperial Chemical Industries) 22, 64, 67, 72–5, 80, 87, 96, 124–5, 129, 134, 138, 146, 155, 192, 195; best hydrocarbon-orientated route 137; carbon efficiency of organisms 27; choice of bacterium as organism 24, 28, 45, 46, 148; converting gas into methanol 52; core businesses 84; 'Deep Shaft' project 158; engineering to sterile standards 139; fermenter 29, 144, 146; Germany 153; huge business opportunities for the future 175; micro-organism investigation 25; new regulatory guidelines for SCP 79; North Seas gas as starting point for bioprotein manufacture 20; novel food 61, 62; oil crises and 69; R&D department 69; reason for methanol for SCP 105; relationship with RHM 77–8, 79; research styles 82–6; screening of organisms 135; search for new markets 56–9; spending on biotechnology 132; toxicity and nutrition tests 151–2; unique low-pressure catalytic process for use of pure cultures 141; value of R&D 127; Zeneca 21; *see also* Marlow Foods; Pruteen; *also ICI divisions below*
ICI Agricultural Division 38, 54–6 *passim*, 104–14, 143, 166; approval for construction of full-scale plant 171; diversification prospects 37; reasons for developing SCP 73
ICI Biological Products (Billingham) 65, 78, 139, 171, 173, 191; development of PHB 158; formation 38; joint project on mycoprotein with ICI 113; process research 83, 101; status and influence of chemical engineers 84
ICI Corporate Lab (Runcorn) 158–9
ICI Educational Publications 213n
ICI Leicester Lab 158
ICI Plant Protection Division (Fernhurst) 5
IFP (Institut Français du Pétrole) 73, 117, 135, 139, 141, 153
Il Sole 64, 122, 214n
immuno-suppressant drugs 125
imports 200, 208, 209, 210
India 210
Indonesia 66
inductive reasoning 11

infection 138, 139, 140, 147, 150; toxic, resistant to 205
information 8; natural inclination not to share 74; optimal choices based on full access to 7
innovation: chemical engineering 4; incremental 5; linked to existing technology, products and markets 183; managers and 4–5; market concept 68–9; radical 5, 17; single-cell protein 17–18; socio-cognitive approach to 180–5, 214–15nn; synoptic model 11, 13; technical, and the economy 1–16, 213n
inoculants 160
inorganic chemistry 89
investment: aggregate pattern of a long wave 185; bunching of 185; capital 137, 187, 204, 205; North Sea 116
Iranian revolution 87
iron 29, 43, 46
isoelectric point 148
Istituto Superiore di Sanita 103
Italproteine 156, 157
Italy 67; anti-SCP consumer movement 103; clash of cultures 118–20; health authorities 79, 104, 118, 119, 121, 122–3, 124, 125, 126; nucleic acid issue a serious problem 154; political sabotaging of SCP plants 193; principal cause of failure of SCP 48; scientific debate over SCP; toxicity 124–6; soya lobby 121–2; *see also* ANIC; BP Sarroch; Liquichimica

JACNE (food advisory group) 61
Jamsankoski 198
Japan 17, 38, 67, 118–19; aquaculture 59; Association of Agricultural Cooperatives 194; banning of SCP 122; food flavouring producers 57; Kondratiev waves and technical innovation 3; market rejection of VCR prototypes 14; Ministry of Agriculture 196; Ministry of Health and Welfare 118, 194, 196; MITI 195, 196; nucleic acid market 57; political sabotaging of SCP plants 193; principal failure of SCP in 48; strong consumer lobbying against SCP 215n; unreliability as long-term customers 58; *see also* Dainippon; Kanegafuchi; Kiro Hako; Kyowa Hakko Kogyo; Mitsui; Nippon-Soda; Subito
John Brown Engineering 38, 55–6, 66, 107, 133, 145, 147; Pruteen construction team 65, 139, 145, 159
joint ventures 116, 121

Kanegafuchi Chemical Corporation 65, 79, 102, 118, 125, 193–4, 195, 196
Karina B 123
Kay, N. 6
Keilling, Prof. 205
Kellogg 62
Kennomeat meaty chunks 118
Kirishi 188, 189, 190, 193
Kiro Hako 72, 115
Knorr, D. 132
knowledge 164; base 5, 162; basic science 75; definition of technology in terms of 165; formal 69; physical artefacts embody 163; selection of 180; socially conditioned 163; socially structured, technology as a form of 181; technical 5, 6, 163, 184
Kondratiev waves 1–5, 185
Korea 57
Kraft process 199, 200
Krasnodar 188
Kuhn, T.S. 11
Kuwait 66, 67, 117, 155
Kyowa Hakko Kogyo Co. 196

lactic acid 205
lactose 25, 205, 207
lactoserum 204, 205
Laine, B.M. 49, 168
Langrish, J. 14
Lavera *see* BP France
Leicester University 158
Leningrad 190

'Liason Council to Ban Petroleum
  Protein' 194
Libya 37, 66, 114, 116, 169
licensing arrangements 195
lignin 26, 197
lignin sulphate 197, 199
lignocellulose 197
lignose sulphates 200
Lilly, M.D. 159
Lindblom, M. 35n
lipids 161
Liquichimica 21, 22, 45, 65, 74, 87,
  121–4 *passim*, 193, 194; arguing
  for approval for process 79–80;
  bankruptcy 126; choice of stir-tank
  fermenter 144; main events 39;
  management and government
  incentives 101–4; marketing
  damaging to SCP project 64;
  media distrust of toxicological
  claims 103; pilot plant 23; ready to
  begin production 119; value of
  R&D 127; yeasts 125
Liquigas 101
Liquipron 104, 122, 144, 154
liver necrosis 154
London 120
long waves *see* Kondratiev
lubricants 199
lysine 32, 41, 42

McKinsey 88
MAFF (Ministry of Agriculture,
  Fisheries and Food) 38, 151
magnesium 29
maize 36
Makarios III (Michael Christodoulos
  Mouskos) 208
managers: behaviour of 6–10;
  cognitive process on the part of 68;
  decision-making 15; and
  innovation 4–5; network 15–16;
  see also senior management
Manchester Business School 112
Mansfield, E. 14
March, J.G. 7
market concept 180, 182; affects
  research 48–51; cognitive elements
  which make up 183–4; extent to
  which it influenced evolution of

projects 44–6; ideas of use and
  need in 67–8; two types 68–70
market pull 58
markets 106, 186, 205; alternative
  56; 'applying science' to need 162;
  burger, steaklet and recipe dish
  63; dairy-free products 59; during
  SCP development 40–70, 213n;
  expected size, affects planned size
  of plant 46; failure of village
  technology to find 209–10; failures
  70; health 59; high-value-added
  57, 187, 202, 206; how ideas
  influenced choices in research and
  development 44; human food 58–9,
  73, 99, 204; hydrocarbon 53; idea
  influences choice of substrate and
  organism 45–6; idea influences
  construction of process plant and
  product 46–8; matching of
  technology to need 165; meat 66,
  78; mycoprotein 63; natural gas as
  a fuel 95; niche 50; nucleic acid 57;
  perceived need 182; perception of
  trend leads to product
  development 59–65; peripheral
  food 61; protein 51–6, 200; soya
  176; soya-meal 53; veal calf milk
  replacer 38, 41, 43–4, 46, 56; yeast
  protein 202; *see also* animal feed
  market
Marks & Spencer 42
Marlborough Polymers 158
Marlow Foods 21, 38, 78, 101, 132
Marseilles 168
Marstrand, P.K. 213n
Martigues 115
Mason, M. 7
meat 42, 66, 78; from animals fed on
  Paprine 189–90; substitute 60, 62,
  63
Medawar, Sir Peter 11
Mennonite farmers 209
mercury poisoning 118, 195
Messina (Sicily) 102
metabolism 124, 125, 137, 140, 154
Metcalfe, S. 6
methane 25, 26, 73, 139; bacteria
  consuming 94, 141; fermentation
  172; organisms consuming 137;

processes 136; projects 141;
recycling 136, 142; solubility of
138; very low efficiency 192; *see
also* alkanes
methanol 26, 76, 111, 175, 192;
ability of certain bacteria to grow
on 18; acted as a toxin 141;
ammonia and 148; aqueous
solution of nutrients and 291;
bacterium consuming 28; culture
104; jumbo plants 117; less
refined, technically possible to use
109; mathematical model for ideal
way in which a micro-organism
could thrive on 145; poor mixing
of water and 146; prices 34, 52–3,
54; production control 73;
research work on 74; SCP based
on 34, 105, 187, 195; substrate 25,
37, 66, 142; supplies 99; switching
from paraffins to, as feedstock 99
methionine 32, 41, 42
methylated spirit 26
Methylophilus Methylotrophus 24
Metsa Selva 198, 200
Mexico 66
Mezzogiorno 117, 214n
microbial processes 85, 131, 149,
156, 197; *see also* animal feed;
proteins; technology
microbiology 119, 161, 162, 168, 169,
171; background training in 176;
dearth of possible projects 157;
research 126, 156; skills 97
micro-organisms 23–4, 25, 29, 149,
154, 203; amino acid profiles vary
between strains of 32–3; foreign
209; thriving on methanol,
mathematical model 145; growing
experimental cultures 22–3;
industrial processes for fermenting
18; *see also* bacteria; yeasts
micropolitical process 176, 177, 179
Middle East 117
Milan 120
milk: powdered 62; veal calf replacer
market 38, 41, 43–4, 46, 56, 58,
106, 186, 205, 206
minerals 29, 76, 133, 151
Minimata 118, 195

MIT (Massachusetts Institute of
Technology) 18, 123
MITI *see* Japan
Mitchell, J.C. 11, 12
Mitsui 102
molasses 19, 87, 89, 91, 160;
fermentation of 209
monoclonal antibodies 77
monoseptic operation 76–7, 144, 145
Montebello di Calabria 104
Montpelier University 205
Moses, V. 130
motivations 167, 168, 171, 178, 179
mushrooms 28, 45, 64, 134
mutations 140, 141
mycoprotein 45, 77, 132, 155;
developing market for 63;
fermentation 38; image 64;
marketing 62, 101; origins of
59–61; production 78; projects
82–4, 99, 129, 159–60, 167–8

NACNE (food advisory group) 61
n-alkanes 24–5, 26, 37, 39, 117, 190;
crude oil with a very high content
114; derived proteins 194;
economics of SCP 143; production
192; feedstock derived from crude
oil known to contain carcinogenic
aromatic hydrocarbon compounds
124; fermentations 143; process
115; purified substrate 188;
residual 144
National Federation of Agricultural
Cooperatives 196
natural gas 25, 53, 104, 144, 186,
187; instead of town gas 52; large
reserves 95; liquified 66; major
constituent of 26; market for, as a
fuel 95; price of 38; seeking new
uses for 94; substrates 142, 152;
waste 136–7
Nature 37, 159
negotiation 9, 181
Nestlé 62, 161
Netherlands 152
new entrants 2, 74, 79, 80
New Scientist 95
niche markets 53, 59
'Nimble' 59

Nippon-Soda 102
nitrogen 24–5, 29; inorganic 131;
    oxidised to nitrate 131
no-till farming technique 55
Nobel Prize winners 143
non-sulphur, non-chlorine pulping
    process 204
Norsk Hydro 20
North Sea gas 20, 25, 52, 66, 94, 116
Norway 199, 200
novel foods 10, 21, 32, 61–2, 69, 151,
    183; new opportunities open
    through the development of
    technology 155–62; protein
    production technology 29–31;
    research as a fashion 130–2
nucleic acids 45, 56, 133, 148–9, 161;
    bacterial 148; by-product of
    metabolism 154; extraction 155;
    food flavourings 57; fraction of
    Pruteen 38, 57; high-value-added
    market 57; intake 154; *see also*
    DNA; RNA
nucleotides 161
nutrients 29
nutrition 24, 25, 135, 207; animal 99,
    186; science 47–8, 210; tests 31–3,
    38, 150, 151–2; trend for public to
    become more aware 63

OECD countries 213n
oil 26, 66; price falls 19–20, 67, 113,
    187; enhanced recovery 157;
    specialist lubricant for drilling bits
    199; synthetic materials 4; tertiary
    recovery 131; *see also* crude oil; oil
    price shocks
oil price shocks/rises: first (1973) 20,
    37, 43, 51–2, 54, 69, 103, 115,
    117–18; second (1979) 20, 36, 43,
    69, 110
olefine 102
Olson, S. 132
organic chemistry 89
organisms 19, 74, 76, 131, 158;
    biochemical pathways 159; choice
    of 25–8, 45–6; fibrous 134; foreign
    138, 139; growth of 75, 129;
    methane-consuming 137;
    pathogenic 138; selection of 133–

42, 162; susceptible to infection
    140; thermo-tolerant 137–8; *see
    also* micro-organisms
oxidation 97, 142
oxygen 29, 136, 143, 145, 146, 155;
    sensitive concentration sensors
    149; solubility of 138; *see also*
    BOD

palm oil producers 20
paper production 197
Paprine 189–90
paraffins 94, 101, 102, 114, 115–16,
    192; considered waste product
    102; fermenting 191; purer
    substrates 154, 188; switching to
    methanol as feedstock 99
patents 73–7, 138, 148
pathogens 138
Pavitt, K. 213n
peat 21
Peet, W.A. 49
Pekilo process 23, 197–8, 199, 200,
    202
penicillin 134, 141, 214n
Perez, C. 2, 3n, 4, 5n, 184
Peru 54, 198
petroleum 19, 194; protein products
    45, 64, 102, 104, 122, 194
Pettigrew, A. 16
pH values 24, 140, 148, 155
Phillips Petroleum 138, 158
phosphate 47, 48
phosphoric acid 25
phosphorus 25, 29
physical chemistry 89
pies 77; *see also* 'Savoury Pie'
    (RHM)
pigs 48, 124, 152, 208; piglets 41, 42,
    56, 186
Pilkington 170
pilot plants 23, 27, 38, 39, 72, 78,
    145, 161, 202, 205, 209; cost of 99;
    demonstrating company
    commitment to technology 115;
    need for 90, 101; variation in
    technical complexity 22
Pinch, T.J. 214n
Polanyi, M. 213n
pollution 208, 210; air 193, 200;

236 *Index*

industrial 189, 195, 202;
  regulations 203; water 200
polyethylene 22
polyhydroxybutyrates 95, 118, 158
polymers 4, 131
Popper, Sir Karl 11
potassium 29, 151
potassium hydroxide 151
poultry feed market 56
Pravda 189, 190, 191
Prentis, S. 132
prices 1, 63, 79; animal feed 53, 202;
  cereal 55; crude oil 36, 52;
  feedstock, high 117; fish 54; fish-
  meal 34, 186; hydrocarbon 36,
  51–3, 113, 184, 185; maize 36;
  methanol 34, 52–3, 54; natural gas
  38; paraffin 94; protein 54, 98,
  198, 199, 200; retail 33–4; soya 20,
  21, 56, 109, 110, 117, 199; soya-
  meal 21, 33, 43, 51, 53, 54;
  substrate 33; *see also* oil price
  shocks
Probion 152
product development 38; perception
  of a market trend leads to 59–65
product quality 14
production technology 163–6; firm
  defined by 182, 184; novel protein
  29–31; pulp 199; radically different
  71; trouble with operation 191
profits 3, 77, 105, 116, 137, 178;
  monopoly 2
Project SAPPHO (1972) 179
propane 97, 142
propionic acid 124–5
protein 17, 21, 46, 76; alternative
  sources 41–2, 100; amino acid
  profile 31; animal feed 18, 148,
  193, 208; bacterial 27, 28, 31, 58,
  148; cell 148; coagulated 27, 148;
  concentrates 205; crisis 175; dairy
  wastes 204–7; demand outstripped
  supply 175; fermentation 26;
  forecasting the market 51–6; fungi
  28; human food 27, 148, 204, 205;
  insufficiency of 210; isolates 43;
  microbial 33–9, 188; n-alkane
  derived 194; new method of
  producing 64–5; novel food/feed/

technology 29–31, 32, 130, 135,
  148, 161; nucleic 23; petroleum-
  derived 45, 64, 102, 104, 122, 194;
  plant 131; potential new sources
  40; prices 54, 98, 198, 199, 200;
  safe product 144; shortage in many
  Third World diets 20; shortages
  55, 168; soluble in water 148;
  substituted for soya- and fish-meal
  32; supplements 18, 19, 38, 87;
  synthetic 98; yeast 28, 202, 204;
  see also mycoprotein; protein gap;
  Pruteen; SCP
protein gap 20, 40, 53–6, 79, 98, 207;
  bridging 64
Protibel 58
Pruteen 29–39 *passim*, 76, 99, 105,
  106, 108, 109, 135, 187; attempt to
  modify process to make human
  food 86; base for future products
  157–8; chemical engineers and 84;
  considered using for human food
  48–9; desirable base technology
  for the future using methanol 110;
  estimates of economics 54;
  fermenter 145, 149, 159;
  government approval for 123;
  granular form of 46–7; hunt for
  alternative markets 56;
  hydrolysing proteins with enzymes
  148; in fish diets 152; lost
  commitment to 78; negotiating for
  supplementary approval 153;
  nucleic acid fraction of 38, 57; only
  realistic use for 58; personnel
  changes at end of 133; plant
  construction 65, 107; potassium
  deficiency 151; problem with 154;
  project champions 166, 167, 171,
  173, 174, 175, 171, 214n;
  proteinaceous fraction of 56–7, 58;
  reorganisation and end of 110–14;
  selenium deficiency 69; spending
  on 132, 150; technology worth
  supporting for its own sake 107;
  transfer of technology 191;
  uneconomic to produce 54, 161;
  value to compounders 42
Pseudomonas 24
public protest groups 195

Queen's Award for Industry 14
'Quorn' 21, 38

R&D (research and development) 7,
    60, 97, 129, 174, 180–2 *passim*,
    197, 202, 208; abandoned 20;
    achievement 143; biological 85;
    'commandments' for 14;
    commercial scrutiny in 93–4;
    control of 91–3; decision-making
    in relation to projects 176;
    differences in spending between
    firms 6; diversification through
    87–90, 94; how market ideas
    influenced choices in 44;
    innovations conceived and
    developed in 4; relationship
    between company and 127–8;
    relations between management
    and 95, 98; role market ideas play
    in 50–1; site where basic science
    'applied' to produce new
    technology 13; skills, exchange of
    fermentation skills for 72; strategy
    204; value of, in different
    companies 127
Rabin, B. 130
Rank, Lord 38, 59, 99, 167–8
rational behaviour 7
RCA (Radio Corporation of
    America) 13
Reggio di Calabria 39, 102, 104
Reynolds, Sir Peter 99
RHM (Rank Hovis McDougall) 54,
    61, 73, 74, 106–7, 152, 167–8, 170;
    change from penicillin to
    Fusarium 38, 76, 214n; choice of
    organism influencing 45; cost of
    toxicological and nutrition tests
    150; Hoechst's Schlingmann
    process and 149; joint project on
    mycoprotein with ICI 113; micro-
    organism investigation 25; need
    for texture in product 134, 155;
    organisms referred to by numbers
    76; process control 159–60; R&D
    team's achievement 143; reduced-
    starch health bread 59;
    relationship with Du Pont 86–7;
    relationship with ICI 77–8, 79;

research styles 82–6; 'Savoury Pie'
    product 38, 60, 63, 64; screening
    of organisms 135; senior
    management decisions 99–101;
    testing procedure 151;
    understanding of the public
    perception of healthy foods 62–4;
    use of pure cultures 141; value of
    R&D 127; *see also* Marlow Foods
rice husks 192
Rimmington, Anthony 188, 190,
    191, 192–3
Riviere, J. 140
RNA (ribonucleic acid) 23–4, 27,
    135, 154
Romantschuk, H. 198n, 199
Rorschach inkblots 12
Rosenbloom, R.S. 13–14
Rothman, H. 146
Rothwell, R. 213n
Royal Dutch Fermentation
    Industries 96; *see also* Shell
rules of thumb 165, 183; *see also*
    heuristics
Rushton stirred tank 144
Russia *see* Soviet Union

Sable sur Sarthe 205
Sainsbury's 38, 78
salts 29, 62, 133
Sardinia *see* BP Sarroch
Sassiyen 58–9, 205, 206
Saudi Arabia 66, 67
'Savoury Pie' 38, 60, 63, 64
Scandinavia 204
Schlingmann process 148–9, 154–5
Schumpeter, J.A. 2, 4, 178, 213n
science 13; applying, to a market
    need 162; bandwagon effect 130,
    131, 132; basic knowledge 75;
    debate over SCP toxicity 124–6;
    food 85–6; frontier-breaking,
    front-rank 143; fundamental 83;
    informal network 132–3; method
    12; nutrition 47–8 suspicion of
    mixed cultures 141; *see also*
    biological sciences; microbiology
SCP (single-cell protein) 39, 71–4,
    86–8, 92–3, 101–5, 118–27, 129–33,
    172–3, 181, 184–5, 214n; alkane

programme 65; amino acid profiles
32; animal feed market 40–1, 71,
99; banning 122; carob project 90,
170; contaminated with
carcinogenic material 118–19;
development 40–70, 196, 213n;
failure 48, 156; father of 169;
fermentation plants 117; fungal
132; government safety tests on
195; human food 99, 154;
hydrocarbon-based 20, 23, 33, 41,
59, 95, 136; initial attraction of
114; innovation 17–18; licensing
65–7, 104; markets and
environment 40–70, 147, 213n;
methanol-based 34, 105, 109, 187,
195; most suitable substrate for
104; n-alkane product 190;
natural-gas-based process 144,
187; non-aseptic processes 140;
organisms synthesised protein
from simple nitrogen compounds
131; origin of the idea, conflicting
accounts 169; paraffin-based
technology 115–16; petfood 195;
plant dedicated to making 33;
political sabotaging of plants 193;
principal bioindustry 189;
production 40–1, 104, 142, 160–1;
pure substrate desirable for
process 153; pursuit of commercial
process 155; research 171;
response of animals to high
percentage diets 154; science
fiction image 131; short history of
18–22; taint of 173; technology 19,
65–7, 115–16, 117, 157; testing
150, 194–5; Third World 207–10;
toxicity 124–6, 193; value to
compounders 42; wastes 196–207;
world's greatest producer 188–93;
yeast feeds 150; *see also* Pruteen;
Toprina
Scrimshaw, Nevin 18
scrubbers 149, 188–9
Second World War 18, 19, 131
selenium 47–8, 69
Seletectum 76
Senez 215n
senior management 82–128, 181, 214n

sensory perception 9
separation 147, 191, 205
serum 205
shake flasks 23, 25, 27
Sharp, David 126, 214n
Shell 53, 55, 66–7, 71, 85, 86, 114,
117, 119, 132, 135–6, 138–43
passim, 155, 192; biological
science research 171–2; Chemicals
Division 94, 95; core business 96;
Corporate Research 94, 95; early
and inexpensive failure of SCP
156; fermenters 144; Natural Gas
Division 94, 95; problem with
Pruteen 154; reason for
involvement in SCP 73; SCP
project and senior management
94–8; value of R&D 127
Sherwood, M. 150
Sieber, Dr 174, 204
Silver, S. 132
Simon, Herbert 213n
Sinskey, A.J. 132
Sittingbourne 94, 97, 171, 172
skimmed milk 47, 58; alternative
uses for 43; powdered 43–4, 58,
106, 110, 186; replacer market 38,
44, 46
slurries 21–2, 28, 31
Snell, R. C. 49, 168
socio-cognitive process 162–3, 163,
180–5, 214–15nn
sodium 29
sodium caseinate 62
sodium chloride 62
sodium citrate 104
sodium hydroxide 151
Solomons, G.L. 18, 19, 22, 130
Soviet Union 17, 18–19, 65, 66, 117,
125, 188–93
soya 66, 103, 121–2, 125, 126, 150;
consumption of imported feeds
200; meat-substitute fiasco 62, 63;
price 20, 56, 109, 110, 117, 199;
production 55, 56; understanding
of the market 176
soya-meal 20, 34, 41, 42, 71;
evolution of market for, as animal
feed 51; import substitition 210;
imports 208; prices 21, 33, 43, 51,

53, 54; protein product
progressively substituted for 32
Spain 116
Spencer, C. 21
Spicer, Arnold 99, 135, 168
spices 161
Spillers 41, 118
spin-off products 158, 161
Spinks, A. 99
spinning process 161
spray-drying 31, 149, 185
Stalingrad 19
Stanley, R. 40
starch 20, 38, 59
steady state reactors 162
Steingass, H. 153
Steinkraus, K. 197
sterilisation 139, 147; electrodes
resistant to 155
sterility 28, 135, 138, 140, 148
stirred tanks 143, 144
stomach and bowel disorders 190
Strauss, A. L. 16
Subito 102
subsidies 56, 117
substrates 21–2, 27, 36, 99, 145, 160;
alternative 191; carbon 28; carob
bean husk 170; cheap 18, 20;
control over supply 162; crude oil
154; doubts about using 153;
experimental 192; fermentation of
20, 90; food-grade 45; fungi can
grow on a range of 135; gas–oil
188; gaseous 136; glucose 21; how
to choose 24–5; hydrocarbon 19,
20, 36, 51, 194; lactose 207;
market influences choice of 45–6;
methanol as 37, 66, 142; most
suitable for SCP 104; n-alkane
188; natural gas 142, 192; non-oil
derived 20; paraffin 154, 188; price
of 33, 34, 51; principal 25, 26;
pure, desirable for SCP process
153; solid 205; soya-meal 51;
testing a variety for fermentation
suitability 90; toxicity of 153–5;
undigested alkane 144; upon
which micro-organisms grew with
difficulty 138–9; waste 28; water-
insoluble 19

sucrose 50, 208
Sudan 160
SugarCo 16, 29, 54, 74, 130, 138,
170, 207–10; *Aspergillus Niger*
fungus 135, 161–2, 208, 209; carob
project 173; control of research
91–3; decision to build and site
SCP pilot plant 90–1; expenditure
22; Hoechst's Schlingmann
process and 149; hole-in-the-
ground fermenter 51, 68; 'New
Ventures' 92, 93; research
programme 49; role of strategy
and senior management 87–94;
Technical Services department 50;
value of R&D 127
sugars 19, 62, 160, 197, 202; hexose
198; pentose 198, 202; refining 87;
removing, from lignose sulphates
200
sulphite 198, 199, 200, 202, 203;
liquor 19, 26, 197, 199, 200
sulphur 25, 41
sulphur dioxide 200, 202, 204
sulphuric acid 25, 197
supplements 18, 19, 38, 59, 87, 206
supply–demand balance 55, 175
surfactants 157
Sweden 3
Switzerland 101, 200, 204
synthetic sweeteners 93

Talin 94
technical choices 51; in the
development of the novel
fermentation projects 129–65,
214n
technical problems 115, 135
techno-economic paradigm 4, 5, 6
technology 5–6, 17–39, 61, 65–7, 94,
117; base 100–1; based on
hydrocarbons 73; before market
49–50; carob fermentation 90;
centrifuging 148, 185;
commitment to 115; control of 74;
core processing, abandonment of
204; definition, in terms of
knowledge 165, 181; developing
130–1, 155–62, 184; disreputable
122; exchange of 149; extremely

important 107; fabrication 147; matching to market need 165; microbial 37; mills closed or superseded by 57; new 8, 13, 57, 73, 162, 185; novel foods 155–62; paraffin-based 115–16; primitive 192; process 80; prototype, giving physical form to 68; pulp-processing 204; spray-drier 185; suggested Third World application (SCP) 160; synthetic fatty acid 102; village 29, 73, 209–10; *see also* fermentation technology; production technology
technology-push 51, 93
Teece, D.J. 5, 182, 184, 213n
texture 45, 62, 72, 101, 134, 135, 161
Thailand 90
Third World 18, 40, 41, 86, 207–10; low-cost, alternative source of protein for 100; nutrition product 134; problems of having populations accept a new source of food 210; protein supplement 38; shortage of protein in many diets 20; suggested application of SCP technology 160
Thomas, H. 7
thrush 125
Tikhonov, Nikolai 192
'Tokyo Liaison Council' 194
Toprina (BP's SCP product) 43, 49, 73, 154, 168, 177; advertising 62; BP Board and 114–26; fashioning a company image through selling 64–5
TOTAL 73
town gas 52, 94, 105
toxicity/toxicology 121, 122–3, 145, 150–5, 193; compounds 197; debate over 124–6, 144; effects 32; formic acid 141; monitoring 139; objections 119; research 194; resistant to infection 205; sensitivity to possible accusations 48; testing 31–3, 151–2, 188; validity of trials 122
trace elements 47
Trading Standards Board 58
trustworthiness 122, 124

turkey meat 42
Turner, B.A. 181

Ufa 188, 192
Uhde 98, 99, 160
Unilever 60, 62, 77
United Kingdom 206; academics extremely free with their ideas 77; conversion from town gas to natural gas 94, 105; food safety 79, 149; government financed project to grow Candida Utilis 19; Kondratiev waves and technical innovation 3
United Paper Mills 198
United States 19, 41, 55, 125, 206; consumer electronics industry 13–14; Kondratiev waves and technical innovation 3; soya 54, 56
University of Hohenheim 153
University of Odense 132
UNPAG (UN Proteins Advisory Group) 79, 154, 155, 214n
Upper Nile 160
urea 131, 208
uric acid 24, 154
Ursini, Dr 101, 102, 103
USSR *see* Soviet Union

value for money 63
VCR products 14
Vendôme 205
Venezuela 37, 66, 117
Veterinary School of Munich 152
village technology 29, 73, 209–10
Villasen 131, 132
viscosity-increasing agents 161
vitamins 59, 205, 206
Volga-Urals fields 188

Walker, Martin 189
wastes 23, 135, 210; biological disposal 169; carbohydrate 18, 20, 50, 196, 197, 199; carbon dioxide 29; citrus 49, 91, 209; feedstock 136; fodder yeast from 33; heavy fraction of paraffins considered 102; lactic process 59; lignin-related 26; liquid, disposal 89; liquor 202; means to solve disposal

problems 20; natural gas 136–7; pulp 19, 57, 197–204; SCP from 196–207; sewage 197, 203; slurries 21–2, 31; vegetable processing 21; yeast 28, 59, 207

water 19, 26, 197, 198, 207; pollution 200; poor mixing of methanol and 146; proteins soluble in 148; waste 20, 202

Weick, Karl 8, 9, 15, 177, 181

West Indies 19

wheat 210

Winkofsky, E. P. 7

wood pulp 20, 26, 196, 197–204

yeast 6–7, 23, 25, 45, 139, 191; animals fed on 190, 194; biomass 19, 59, 197; brewery 206; cells 124, 126, 148; dust scrubbing equipment 188–9, 190; fermentation 144, 199; flocculated and filtered 205; fodder 33, 58, 198, 199, 202; food supplements 19, 59; good quality 160; grown on alkanes 80, 114; harvesting process 205; human food 37, 204, 206; lactic 206, 207; pathogenic 103, 125, 126; powdered 59, 206; processes 147–8; products 58; protein 19, 28, 199, 202, 204; research into genetic structure of 159; researching growth of, on hydrocarbons 193; Saccharomyces 199; strains 205; that causes thrush 125; used to bind proteins in sausages 207; vitamin preparation for shiny dog coats 59, 206; *see also Candida*

yields 136, 142, 155, 158, 162

Zegveld, W. 213n

zinc 150

Zurich Polytechnic 138